Firefighter Fatalities
in the United States in 2011

Prepared by

U.S. Department of Homeland Security

Federal Emergency Management Agency

U.S. Fire Administration

National Fire Data Center

and

The National Fallen Firefighters Foundation

www.firehero.org

In memory of all firefighters

who answered their last call in 2011

To their families and friends

To their service and sacrifice

Cover photo courtesy of Mark A. Whitney

Table of Contents

Acknowledgements 1
Background. 1
Introduction 2
   Who is a Firefighter? 2
   What Constitutes an Onduty Fatality? 2
   Sources of Initial Notification. 3
   Procedure for Including a Fatality in the Study 3
2011 Findings 4
   Career, Volunteer, and Wildland Agency Deaths. 6
   Gender. 6
   Multiple Firefighter Fatality Incidents 6
   Wildland Firefighting Deaths 7
Type of Duty. 9
   Fireground Operations. 10
   Type of Fireground Activity 10
   Fixed Property Use for Structural Firefighting Deaths. 11
   Responding/Returning. 11
   Training. 12
   Nonfire Emergencies 12
   After an Incident 12
Cause of Fatal Injury 13
   Stress or Overexertion 13
   Vehicle Crashes 14
   Lost or Disoriented. 15
   Caught or Trapped 15
   Collapse. 16
   Struck by Object. 16
   Fall. 17
   Out of Air 18
   Other. 18
Nature of Fatal Injury 19
Firefighter Ages 20
Deaths by Time of Injury 21
Firefighter Fatality Incidents by Month of Year. 21
State and Region. 22
Analysis of Urban/Rural/Suburban Patterns in Firefighter Fatalities 26
Appendix A 27
Appendix B 65

Acknowledgements

This study of firefighter fatalities would not have been possible without the cooperation and assistance of many members of the fire service across the United States. Members of individual fire departments, chief fire officers, wildland fire service organizations such as the U.S. Forest Service (USFS), the National Park Service (NPS), the Bureau of Land Management (BLM), the Bureau of Indian Affairs (BIA), the U.S. Fish and Wildlife Service (FWS), as well as the U.S. Department of Justice (DOJ), the National Fire Protection Association (NFPA), and many others contributed important information to this report.

The National Fallen Firefighters Foundation (NFFF) was responsible for compilation of a large portion of the data used in this report and the incident narrative summaries found in Appendix A. Their cooperation and work toward reducing firefighter deaths is gratefully acknowledged.

The ultimate objective of this effort is to reduce the number of firefighter deaths through an increased awareness and understanding of their causes and how they can be prevented. Firefighting, rescue, and other types of emergency operations are essential activities in an inherently dangerous profession, and unfortunate tragedies do occur. These are the risks all firefighters accept every time they respond to an emergency incident. However, the risks can be greatly reduced through efforts to improve training, emergency scene operations, and firefighter health and safety initiatives.

Background

For 35 years, the U.S. Fire Administration (USFA) has tracked the number of firefighter fatalities and conducted an annual analysis. Through the collection of information on the causes of firefighter deaths, USFA is able to focus on specific problems and direct efforts toward finding solutions to reduce the number of firefighter fatalities in the future. This information is also used to measure the effectiveness of current programs directed toward firefighter health and safety.

Several programs have been funded by USFA in response to this annual report. For example, USFA has sponsored significant work in the areas of general emergency vehicle operations safety, fire department tanker/tender operations safety, firefighter incident scene rehabilitation, and roadside incident safety. The data developed for this report are also widely used in other firefighter fatality prevention efforts.

In addition to the analysis, USFA, working in partnership with the National Fallen Firefighters Foundation (NFFF), develops a list of all onduty firefighter fatalities and associated documentation each year. If certain criteria are met, the fallen firefighter's next of kin, as well as members of the individual's fire department, are invited to the annual National Fallen Firefighters Memorial Weekend Service. The service is held at the National Emergency Training Center (NETC) in Emmitsburg, MD, during Fire Prevention Week in October of each year. Additional information regarding the Memorial Service can be found at www.firehero.org or by calling NFFF at (301) 447-1365.

Other resources and information regarding firefighter fatalities, including current fatality notices, the National Fallen Firefighters Memorial database, and links to the Public Safety Officers' Benefit (PSOB) Program, can be found at http://www.usfa.fema.gov/fireservice/fatalities/

Introduction

This report continues a series of annual studies by the U.S. Fire Administration (USFA) of onduty firefighter fatalities in the United States.

The specific objective of this study is to identify all onduty firefighter fatalities that occurred in the United States and its protectorates in 2011 and to analyze the circumstances surrounding each occurrence. The study is intended to help identify approaches that could reduce the number of firefighter deaths in future years.

Who is a Firefighter?

For the purpose of this study, the term firefighter covers all members of organized fire departments with assigned fire suppression duties in all 50 States, the District of Columbia, and the Territories of Puerto Rico, the Virgin Islands, American Samoa, the Commonwealth of the Northern Mariana Islands, and Guam. It includes career and volunteer firefighters; full-time public safety officers acting as firefighters; fire police; State, territory, and Federal government fire service personnel, including wildland firefighters; and privately employed firefighters, including employees of contract fire departments and trained members of industrial fire brigades, whether full or part time. It also includes contract personnel working as firefighters or assigned to work in direct support of fire service organizations (i.e., air-tanker crews).

Under this definition, the study includes not only local and municipal firefighters, but also seasonal and full-time employees of the U.S. Forest Service (USFS), the National Park Service (NPS), the Bureau of Land Management (BLM), the Bureau of Indian Affairs (BIA), the U.S. Fish and Wildlife Service (FWS), and State wildland agencies. The definition also includes prison inmates serving on firefighting crews; firefighters employed by other governmental agencies, such as the U.S. Department of Energy (DOE); military personnel performing assigned fire suppression activities; and civilian firefighters working at military installations.

What Constitutes an Onduty Fatality?

Onduty fatalities include any injury or illness sustained while on duty that proves fatal. The term "on duty" refers to being involved in operations at the scene of an emergency, whether it is a fire or nonfire incident; responding to or returning from an incident; performing other officially assigned duties such as training, maintenance, public education, inspection, investigations, court testimony, and fundraising; and being on call, under orders, or on standby duty except at the individual's home or place of business. An individual who experiences a heart attack or other fatal injury at home while he or she prepares to respond to an emergency is considered on duty when the response begins. A firefighter who becomes ill while performing fire department duties and suffers a heart attack shortly after arriving home or at another location may be considered on duty since the inception of the heart attack occurred while the firefighter was on duty.

On December 15, 2003, the President of the United States signed into law the Hometown Heroes Survivors Benefit Act of 2003. After being signed by the President, the Act became Public Law 108-182. The law presumes that a heart attack or stroke are in the line of duty if the firefighter was engaged in nonroutine stressful or strenuous physical activity while on duty and the firefighter becomes ill while on duty or within 24 hours after engaging in such activity. The full text of the law is available at http://frwebgate. access.gpo.gov/cgi-bin/getdoc.cgi?dbname=108_cong_ public_laws&docid=f:publ182.108.pdf

The inclusion criteria for this study have been affected by this change in the law. Previous to December 15, 2003, firefighters who became ill as the result of a heart attack or stroke after going off duty needed to register a complaint of not feeling well while still on duty in order to be included in this study. For firefighter fatalities after December 15, 2003, firefighters will be included in this report if they became ill as the result of a heart attack or stroke within 24 hours of a training activity or emergency response. Firefighters who became ill after going off duty where the activities while on duty were limited to tasks that did not involve physical or mental stress will not be included.

A fatality may be caused directly by an accidental or intentional injury in either emergency or nonemergency circumstances, or it may be attributed to an occupationally related fatal illness. A common example of a fatal illness incurred on duty is a heart attack. Fatalities attributed to occupational illnesses also include a communicable disease contracted while on duty that proved fatal when the disease could be attributed to a documented occupational exposure.

Firefighter fatalities are included in this report even when death is considerably delayed after the original incident. When the incident and the death occur in different years, the analysis counts the fatality as having occurred in the year in which the incident took place.

There is no established mechanism for identifying fatalities that result from illnesses such as cancer that develop over long periods of time and which may be related to occupational exposure to hazardous materials or toxic products of combustion. It has proved to be very difficult to provide a complete evaluation of an occupational illness as a causal factor in firefighter deaths due to the following limitations: the exposure of firefighters to toxic hazards is not sufficiently tracked; the often delayed long-term effects of such toxic hazard exposures; and the exposures firefighters may receive while off duty.

Sources of Initial Notification

As an integral part of its ongoing program to collect and analyze fire data, USFA solicits information on firefighter fatalities directly from the fire service and from a wide range of other sources. These sources include the Public Safety Officers' Benefit (PSOB) Program administered by the U.S. Department of Justice (DOJ), the National Institute for Occupational Safety and Health (NIOSH), the Occupational Safety and Health Administration (OSHA), the Department of Defense (DOD), the National Interagency Fire Center (NIFC), and other Federal agencies.

USFA receives notification of some deaths directly from fire departments, as well as from such fire service organizations as the International Association of Fire Chiefs (IAFC), the International Association of Fire Fighters (IAFF), the National Fire Protection Association (NFPA), the National Volunteer Fire Council (NVFC), State fire marshals, State fire training organizations, other State and local organizations, fire service Internet sites, news services, and fire service publications.

Procedure for Including a Fatality in the Study

In most cases, after notification of a fatal incident, initial telephone contact is made with local authorities by USFA to verify the incident, its location, jurisdiction, and the fire department or agency involved. Further information about the deceased firefighter and the incident may be obtained from the chief of the fire department, designee over the phone, or by other forms of data collection. After basic information is collected, a notice of the firefighter fatality is posted at the National Fallen Firefighters Memorial site in Emmitsburg, MD, the USFA website, and is transmitted by electronic mail to a large list of fire service organizations and fire service members.

Information that is routinely requested from fire departments that have experienced a fatality includes National Fire Incident Reporting System (NFIRS)-1 (incident) and NFIRS-3 (fire service casualty) reports; the fire department's own incident and internal investigation reports; copies of death certificates and autopsy results; special investigative reports; law enforcement reports; photographs and diagrams; and newspaper or media accounts of the incident. Information on the incident may also be gathered from NFPA or NIOSH reports.

After obtaining this information, a determination is made as to whether the death qualifies as an onduty firefighter fatality according to the previously described criteria. With the exception of firefighter deaths after December 15, 2003, the same criteria were used for this study as in previous annual studies. Additional information may be requested by USFA, either through follow-up with the fire department directly, from State vital records offices, or other agencies. The final determination as to whether a fatality qualifies as an onduty death for inclusion in this statistical analysis is made by USFA. The National Fallen Firefighter Foundation's (NFFF's) criteria as a line-of-duty death (LODD) for inclusion in the annual National Fallen Firefighters Memorial Service is made by NFFF.

2011 Findings

Eighty-three firefighters died while on duty in 2011. The 2011 total includes 19 firefighters who died under circumstances as a result of inclusion criteria changes resulting from the Hometown Heroes Survivors Benefit Act of 2003. When not including these fatalities in a trend analysis, the 2011 total of 64 firefighter fatalities was, for the second year in a row, the lowest number of firefighter losses on record—a substantial 11 percent reduction from the 72 such firefighters lost in 2010.

An analysis of multiyear firefighter fatality trends needs to acknowledge the changes from the Hometown Heroes Survivors Benefit Act of 2003; therefore, some graphs and charts either will or will not indicate the Hometown Heroes portion of the total. However, this does not diminish the sacrifices made by any firefighter who dies while on duty or the sacrifices made by his or her family and peers.

Moreover, when conducting multiyear comparisons of firefighter fatalities in this report, the losses that were the result of the attacks on the World Trade Center (WTC) in New York City on September 11, 2001, are sometimes also set apart for illustrative purposes. This action is by no means a minimization of the supreme sacrifice made by these firefighters.

Figure 1. Onduty Firefighter Fatalities (1977–2011).

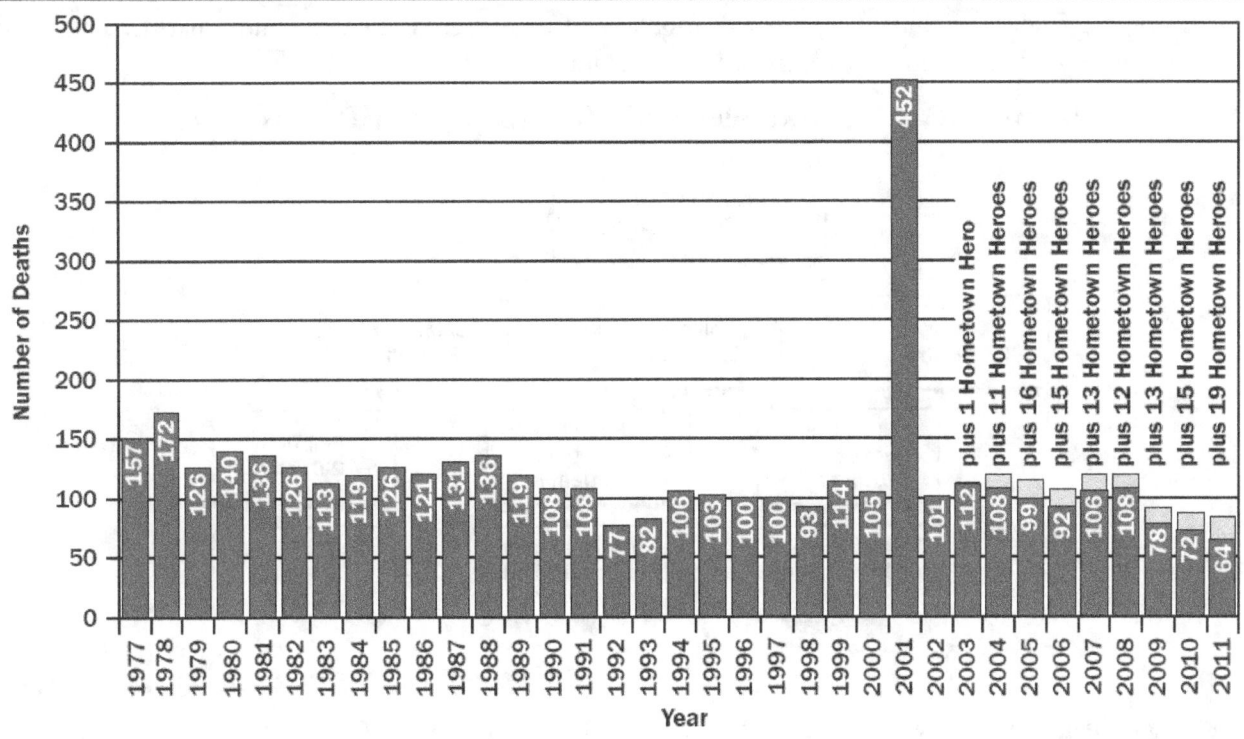

Figure 2. Firefighter Fatalities per 100,000 Fires.

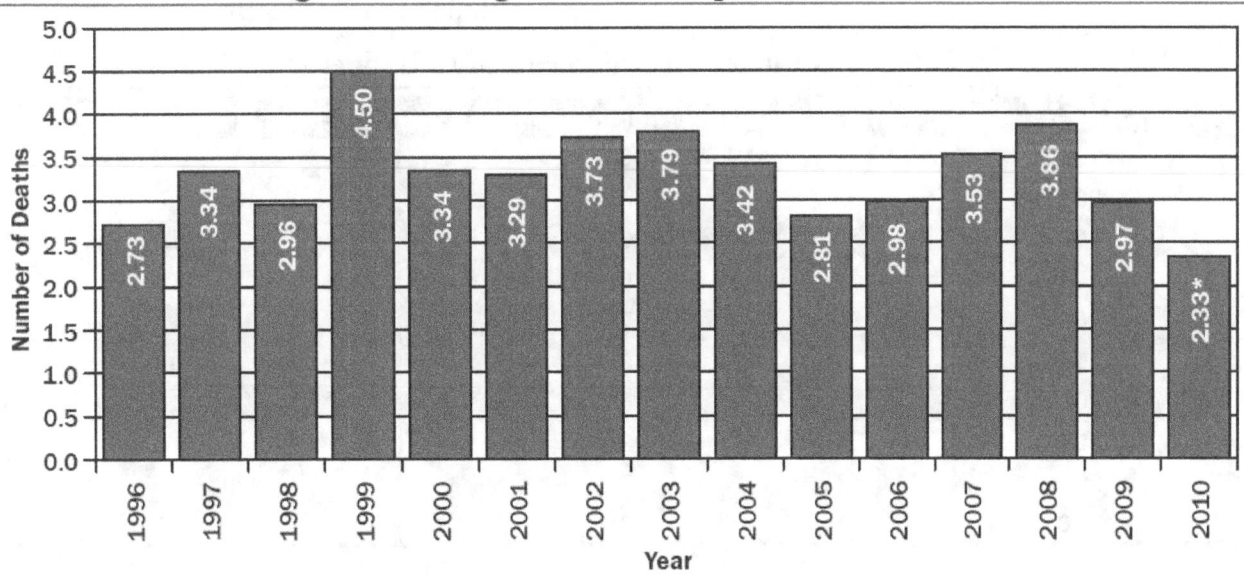

*2011 ratio will be included in the 2012 report.

Career, Volunteer, and Wildland Agency Deaths

In 2011, firefighter fatalities included 27 career firefighters, 51 volunteer firefighters, and 5 part-time or full-time members of wildland or wildland contract fire agencies (Figure 3).

Figure 3. Career, Volunteer, and Wildland Agency Deaths (2011)

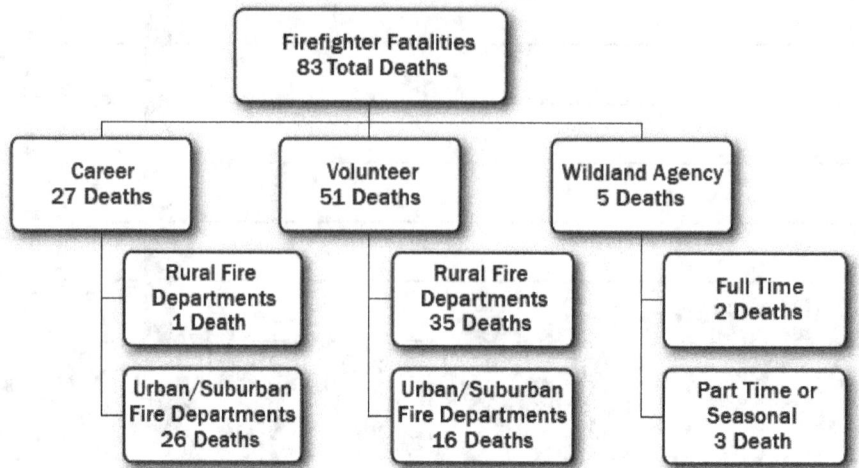

Gender

Of the 83 firefighters who died while on duty in 2011, 82 were male and 1 was female.

Multiple Firefighter Fatality Incidents

The 83 deaths in 2011 resulted from a total of 80 fatal incidents. There were three firefighter fatality incidents where two firefighters were killed in each incident, claiming a total of six firefighters.

Table 1. Multiple Firefighter Fatality Incidents

Year	Number of Incidents	Total Number of Deaths
2011	3	6
2010	4	8
2009	6	13
2008	5	18
2007	7	21
2006	6	17
2005	4	10
2004	3	6
2003	7	20
2002	9	25

Wildland Firefighting Deaths

In 2011, 10 firefighters were killed during activities involving brush, grass, or wildland firefighting. This total includes part-time and seasonal wildland firefighters, full-time wildland firefighters, and municipal or volunteer firefighters whose deaths are related to a wildland fire (Figure 4).

Figure 4. Firefighter Fatalities Related to Wildland Firefighting (2002–2011)

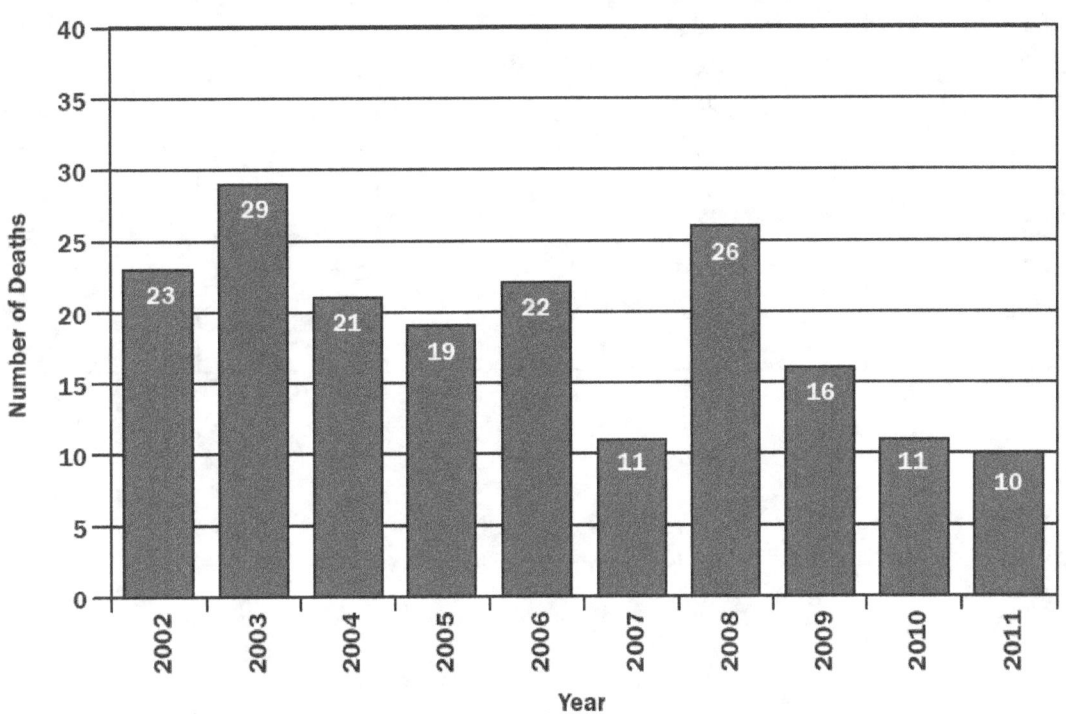

Table 2. Wildland Firefighting Aircraft Deaths

Year	Total Number of Deaths	Number of Fatal Incidents
2011	0	0
2010	0	0
2009	5	3
2008	16	4
2007	1	1
2006	8	3
2005	6	2
2004	3	3
2003	7	4
2002	6	3

In 2011, there was one multiple firefighter fatality incident related to wildland firefighting killing two, and, for the second year in a row, there were no wildland firefighter deaths related to aircraft.

Table 3. Firefighter Deaths Associated with Wildland Firefighting

Year	Total Number of Deaths	Number of Fatal Incidents	Number of Firefighters Killed in Multiple-Death Incidents
2011	10	9	2
2010	11	11	0
2009	16	13	5
2008	26	15	14
2007	11	11	0
2006	22	13	13
2005	19	15	6
2004	21	21	0
2003	30	22	10
2002	23	14	13

Type Of Duty

Activities related to emergency incidents resulted in the deaths of 45 firefighters in 2011 (Figure 5). This includes all firefighters who died responding to an emergency or at an emergency scene, returning from an emergency incident, and during other emergency-related activities. Nonemergency activities accounted for 38 fatalities. Nonemergency duties include training, administrative activities, performing other functions that are not related to an emergency incident, and postincident fatalities where the firefighter does not experience the illness or injury during the emergency.

Figure 5. Firefighter Deaths by Type of Duty (2011)

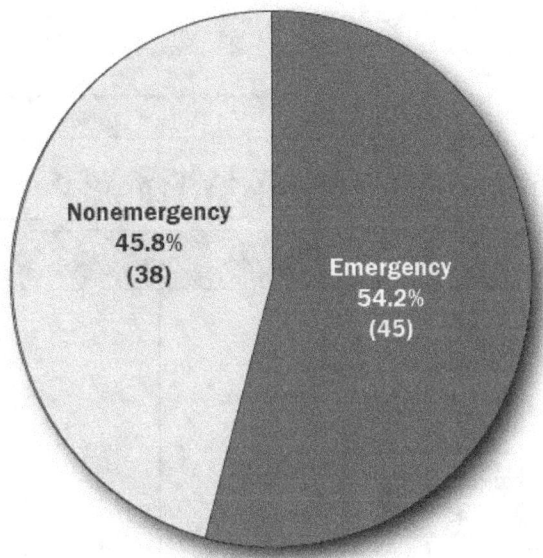

A multiyear historical perspective relating to the percentage of firefighter deaths that occurred during emergency duty is presented in Table 4.

Table 4. Emergency Duty Firefighter Deaths

Year	Percentage of All Deaths	Percentage of All Deaths Without Hometown Heroes
2011	54.2	70.3
2010	55.2	66.7
2009	63.3	82.2
2008	63.5	70.0
2007	64.4	72.4
2006	57.5	66.3
2005	52.1	60.6
2004	68.9	75.9
2003	69.0	69.6
2002	73.0	N/A

The number of deaths by type of duty being performed in 2011 is shown in Table 5 and presented graphically in Figure 6. As has been the case for most years, fire- ground duties are the most common type of duty for firefighters killed while on duty.

Table 5. Firefighter Deaths by Type of Duty (2011)

Type of Duty	Number of Deaths
Responding/Returning	11
Other Onduty Deaths	10
Training	8
After an Incident	21
Fireground Operations	28
Nonfire Emergencies	5
Total	83

Figure 6. Firefighter Deaths by Type of Duty (2011)

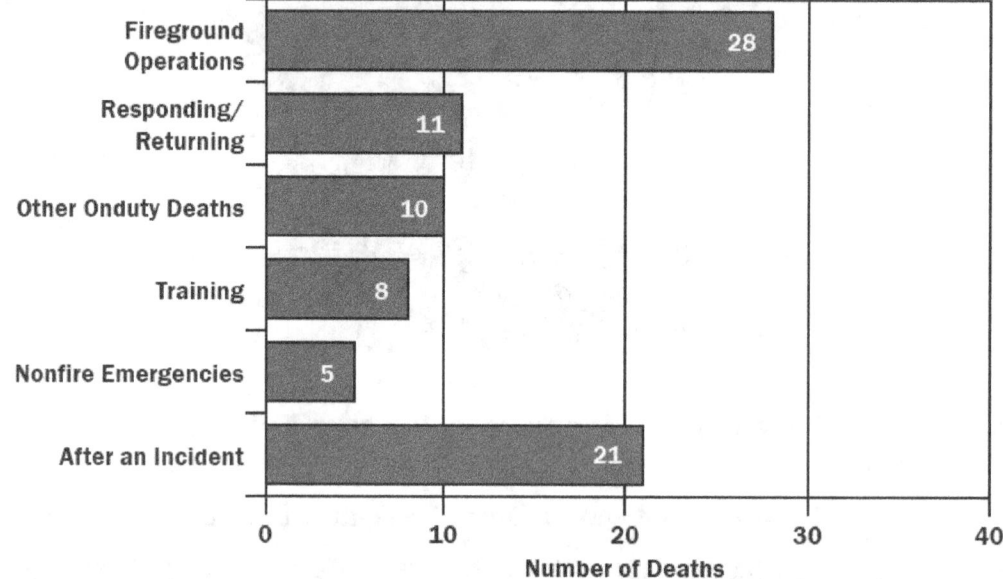

Fireground Operations

Of the 28 firefighters killed during fireground operations in 2011, 21 were at the scene of a structure fire, and 7 others were at the scene of a wildland or outside fire. The average age of the firefighters killed during fireground operations was 43, 13 from volunteer fire departments, 11 career, and 4 wildland (2 each full time and part time).

Type of Fireground Activity

Table 6 shows the types of fireground activities in which firefighters were engaged at the time they sustained their fatal injuries or illnesses. This total includes all firefighting duties, such as wildland firefighting and structural firefighting.

Table 6. Type of Activity (2011)

Pump Ops	1
Search & Rescue	3
Advance Hoselines	14
Water Supply	2
Other	4
Unknown	2
Ventilation	1
Support	1

Fixed Property Use for Structural Firefighting Deaths

There were 21 fatalities in 2011 where firefighters became ill or injured while on the scene of a structure fire. Table 7 shows the distribution of these deaths by fixed property use.

Table 7. Structural Firefighting Deaths by Fixed Property Use in 2011

Residential	14
Commercial	6
Other	1

Responding/Returning

Eleven firefighters died while responding to or returning from 11 emergency incidents in 2011. Nine of the firefighters died while responding to incidents and two, both heart attacks, while returning from an incident.

Four of the nine firefighters killed while responding to incidents died from heart attacks.

Four of the firefighters who died while responding to incidents were killed by trauma caused by motor vehicle collisions, including three in privately-owned vehicles (POVs) and one in a fire department apparatus. A fifth firefighter died from injuries while responding to a carbon monoxide alarm from his residence. He slipped on ice in his driveway, sustained a serious spinal injury, and passed away in the hospital several weeks later.

In the single crash of a fire department apparatus, a 1998 Dodge Utility, taking a firefighter's life in 2011, seat restraints were present but status of use was not reported. The 22-year-old firefighter operating the vehicle was alone in the truck and responding to an emergency medical incident. He lost control of the vehicle in a curve, exited the left side of the roadway, and crashed into an embankment and trees.

Table 8. Firefighter Deaths While Responding to or Returning from an Incident

Year	Number of Firefighter Deaths
2011	11
2010	16
2009	15
2008	24
2007	26
2006	15
2005	22
2004	23
2003	36
2002	13

Training

In 2011, eight firefighters died while engaged in training activities. Five of the deaths were due to heart attacks and one from a cerebrovascular accident (CVA). One firefighter died when he attempted to climb one of two ropes suspended below the raised platform of the department's ladder tower. He likely lost his grip and fell 6 to 8 feet to the ground, sustaining a fatal head injury in the fall.

One Air Force firefighter, acting as a spotter, was crushed between a P-23 Aircraft Rescue Fire Fighting (ARFF) apparatus and a cement column as the vehicle was backed into the apparatus bay.

Table 9. Firefighter Fatalities While Engaged in Training

Year	Number of Firefighter Deaths
2011	8
2010	12
2009	10
2008	12
2007	11
2006	9
2005	14
2004	13
2003	12
2002	11

Nonfire Emergencies

In 2011, there was one fire police and nine firefighter fatalities where the type of emergency duty was not related to a fire.

A fire police officer died from a heart attack while working at a traffic control point providing scene safety for firefighters working a motor vehicle accident (MVA). Two firefighters were also struck and killed at the scene of MVAs while providing scene safety.

One firefighter, a Hotshot crew member, became separated and lost from fellow firefighters after completing their assignment and returning by foot in rough backcountry terrain. His body was discovered the following day, but the cause and circumstances related to his death were not released.

One firefighter died from injuries he sustained when he slipped and fell on ice at his residence while attempting to depart for a response.

Two firefighters died in MVAs while responding to incidents, one in a fire department apparatus and one in a POV. In both instances, they failed to successfully negotiate curved sections of roadway and speed was reported as a contributing factor.

One firefighter responded with his fire company to a technical rescue call but drowned after he entered a cold body of water and began to swim out to rescue two boaters in distress. He became incapacitated due to the cold and subsequently died.

Two firefighters died from heart attacks, one while clearing trees following an outbreak of severe weather and the second after returning home after attempting an Emergency Medical Services (EMS) response for a sick child during a blizzard. The firefighter had become stuck in the snow and after being extricated, returned home and died in his vehicle while in his driveway.

After an Incident

In 2011, 21 firefighters died after the conclusion of their onduty activities. Twenty of the deaths were due to heart attacks, and in one case the cause of death was a pulmonary embolism. Nineteen of the fatalities were classified as Hometown Heroes where no symptom or complaint of illness became evident or were reported during duty. In the two other instances, the firefighters complained of feeling ill while on scene or after returning to quarters and then passed away later from heart attacks.

Cause of Fatal Injury

The term "cause of injury" refers to the action, lack of action, or circumstances that directly resulted in the fatal injury. The term "nature of injury" refers to the medical cause of the fatal injury or illness which is often referred to as the physiological cause of death. A fatal injury is usually the result of a chain of events, the first of which is recorded as the cause.

Figure 7 shows the distribution of deaths by cause of fatal injury or illness in 2011.

Figure 7. Fatalities by Cause of Fatal Injury (2011)

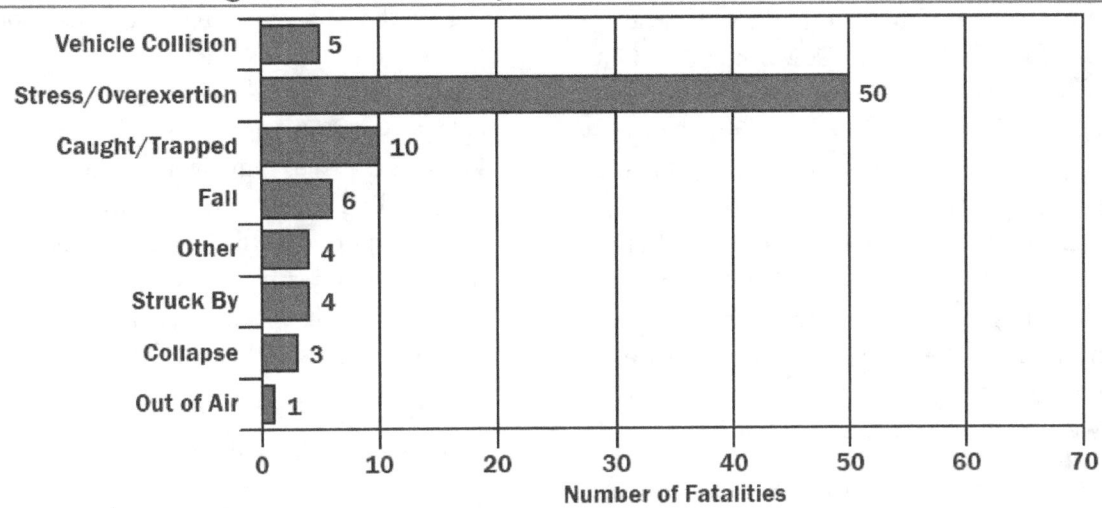

Stress or Overexertion

Firefighting is extremely strenuous physical work and is likely one of the most physically demanding activities that the human body performs.

Stress or overexertion is a general category that includes all firefighter deaths that are cardiac or cerebrovascular in nature such as heart attacks, strokes, and other events such as extreme climatic thermal exposure. Classification of a firefighter fatality in this cause of fatal injury category does not necessarily indicate that a firefighter was in poor physical condition.

Fifty firefighters died in 2011 as a result of stress or overexertion:

- forty-eight firefighters died due to heart attacks;

- one firefighter died due to a cerebrovascular accident (CVA); and

- one firefighter died from heat exhaustion.

Table 10. Deaths Caused by Stress or Overexertion

Year	Number	Percent of Fatalities	Hometown Heroes
2011	50	60.2	18
2010	55	63.2	15
2009	50	54.9	11
2008	54	45.0	12
2007	55	51.4	13
2006	55	53.9	16
2005	62	53.9	15
2004	66	55.5	11
2003	53	46.9*	1
2002	38	37.6	0

*Includes Hometown Heroes, one in December 2003, and an average of 13.9 for the years 2004–2011.

Vehicle Crashes

Five firefighters died in 2011 as the result of vehicle crashes, three involving privately-owned vehicles (POVs) and two involving apparatus: one of the latter as a pedestrian ground guide. The 2011 total reflects a substantial reduction from previous years. Also in 2011, for the second year in a row, there were no firefighter deaths that involved aircraft.

- One firefighter responding to a motor vehicle accident (MVA) lost control of his POV when the right wheels of his pickup left the roadway in a gradual curve. When the firefighter corrected to the left, the truck turned sideways and slid diagonally across the roadway until it began to roll. The firefighter was ejected during the roll and was found 15–20 feet away from his truck where it came to rest. Responding firefighters and emergency personnel pronounced the firefighter dead at the scene from head injuries. Speed and the failure to wear a seatbelt were cited as factors in his death.

- One firefighter was crushed to death while acting as a ground guide for an apparatus that was backing up.

- One firefighter responding to a report of a brush fire lost control of his POV when the right wheels of the Mazda 3 left the right side of the roadway. In response, he corrected to the left and crashed into a minivan headed in the opposite direction. The firefighter was pronounced dead at the scene

of the crash. Law enforcement, reporting on the crash, cited speed as a factor; the status of seatbelts was not reported.

- The youngest firefighter killed in 2011 died while operating a POV, a 2002 Ford Explorer, in response to a mutual-aid structure fire. His mother was riding in the front passenger seat. As the firefighter approached a curve, he encountered a vehicle coming in the opposite direction. His mother stated that the oncoming vehicle was on their side of the road. The firefighter steered to the right to avoid a collision and attempted to correct back to the left but struck a tree, spun around, and struck a second tree, ending back in the roadway. The firefighter was transported to the hospital by medical helicopter but did not survive his injuries. The firefighter was wearing his seatbelt at the time of the crash. His mother received non-life threatening injuries.

- One firefighter died while responding in the department's 1998 Dodge brush truck to an emergency medical incident. He lost control of the vehicle in a curve and was involved in a single vehicle crash. The vehicle exited the left side of the roadway and crashed into an embankment and trees. He was transported to the hospital but later died from his injuries. News reports citing the law enforcement crash report noted that wet roads and speed were factors in the accident. The status of the firefighter's seatbelt was not reported.

Figure 8. Firefighter Fatalities in Vehicle Collisions (Including Aircraft)

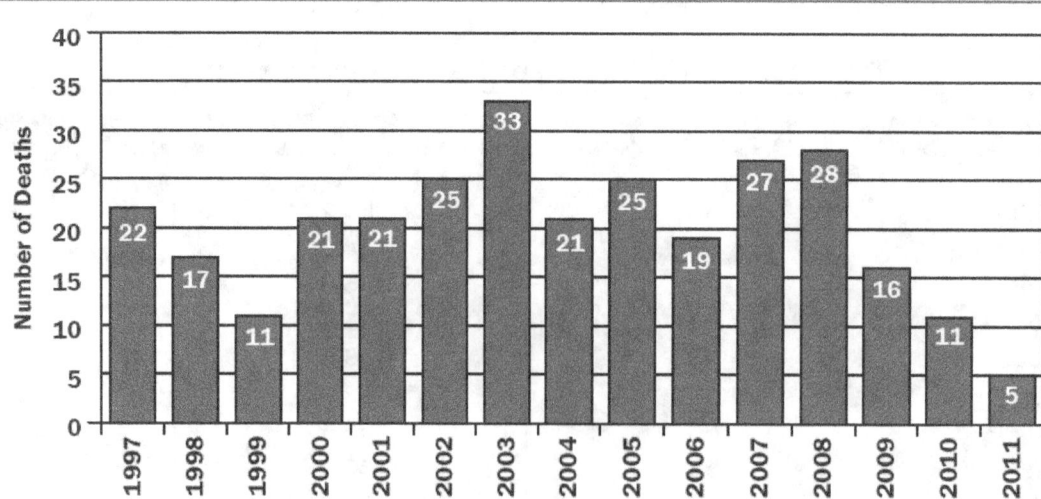

Lost or Disoriented

No firefighters died in 2011 by becoming lost or disoriented inside of a burning structure.

Caught or Trapped

Ten firefighters were killed in 2011 in eight separate incidents when they became caught or trapped; twice the number of firefighters killed in 2010 from being caught or trapped. This classification covers firefighters trapped in wildland and structural fires who were unable to escape due to rapid fire progression and the byproducts of smoke, heat, toxic gases, and flame. This classification also includes firefighters who drowned and those who were trapped and crushed.

- Rapid fire progress trapped and killed one firefighter in a third-story apartment. The Incident Commander (IC) observed the advancement of the fire and ordered an evacuation. Tones were sounded over the radio, and air horns on fire apparatus were sounded. One other firefighter searching the same apartment was able to make it to a window and down a ladder. The firefighter/paramedic declared a Mayday and advised the IC that he was trapped on the third floor.

- One firefighter was killed while working a response assignment to a structure fire in a large (12,500 square foot) residence. Firefighters found fire and smoke showing upon their arrival. Further investigation revealed fire in the walls of the structure that had spread to a large void ceiling space. Firefighters advanced hoselines into the structure, opened the roof, and began to open walls and ceilings on the interior. The interior ceilings were high, and longer pike poles and ground ladders were used. Fire appeared to be running through the attic, and firefighters struggled to expose all areas of fire. A significant portion of the drop ceiling on the first floor of the structure collapsed onto interior firefighters. Three firefighters, including the firefighter killed during the incident, were trapped in or under the debris.

- One firefighter drowned during a response to a local pond for a report of a person in the water in distress. The firefighter entered the water and swam for approximately 20 feet until he became incapacitated, called for help, disappeared underwater, briefly reappeared, and went underwater again. A civilian entered the water and removed the firefighter. He was treated in an ambulance while en route to the hospital and also treated at the emergency room. He was pronounced dead at the hospital due to drowning. The boater who had been in distress also died.

- One firefighter died while working a mutual-aid wildland fire after two brush trucks became stuck in deep sand. As the fire advanced toward the two incapacitated trucks, several firefighters abandoned their positions and tried to outrun the fire. After the fire burned over the area, the firefighter was found suffering from severe burns. He was transported to the hospital by ambulance but died as a result of his burns.

- Two firefighters were dispatched as part of a larger assignment to a report of a structure fire in a residence. Their engine was the first unit on the scene and reported light smoke showing from the garage. The two firefighters entered the structure to assess conditions. The firefighters stretched a 200-foot preconnected attack line into the structure. Both firefighters wore full structural firefighting protective clothing and a self-contained breathing apparatus (SCBA). As firefighters worked on the scene, a window in the fireroom failed, allowing the introduction of additional oxygen to the fire. The fire progressed rapidly up a stairwell and overcame the two firefighters, killing them.

- One firefighter was killed during a fire fight in a large church. First-arriving firefighters reported visible flames and heavy smoke coming from the roof. The first-arriving Company Officer (CO) called for a second alarm and shortly thereafter called for additional tankers (tenders). Firefighters entered the structure and found mostly clear conditions in the interior of the church. As firefighters began to open up the ceilings to access the attic space, they discovered a considerable amount of fire. Interior crews were having difficulty controlling the fire with a handline. Water supply was an issue; the area of the church did not have fire hydrants. Interior firefighters notified the IC that they were withdrawing from the structure. As firefighters were preparing to leave, a structural collapse occurred. An accountability check was conducted, and it was realized that one firefighter was missing. Due to the volume of fire, firefighters were unable to access the collapsed area.

- Two forest rangers were assigned to a lightning-caused wildfire. The fire had initially been contained, but several days later it had spread beyond the containment area where both forest rangers operated tractor/plow units. During operations, one of the units became stuck. The second unit and operator attempted to provide assistance. As the wildfire approached, both operators abandoned their units and attempted to outrun the fire. They were overtaken by the progress of the fire. Neither firefighter attempted to deploy a fire shelter. Both firefighters died of burns.

- An engine crew was assigned fire suppression duties on a midslope road along the fire edge of a wildfire. The fire rapidly advanced, cutting off the escape route for the firefighters. Two firefighters were able to run to safety, sustaining moderate to severe burn injuries. A third firefighter could not escape from the cab of the engine and died due to smoke inhalation and thermal burns.

Collapse

Three firefighters died in 2011 as a result of structural collapses.

- Firefighters were dispatched to respond to a working fire in a 96-year-old brick and masonry structure that housed an antique store and living quarters. The fire fight was in a defensive mode with no interior firefighting operations. The fire had self-vented and cracks were visible in at least one exterior wall. It was noted that an extension ladder was against the building in an area that would be subject to collapse. Two firefighters entered the collapse zone to retrieve the ladder. One of the firefighters took a position under the ladder, supporting the ladder, as the other firefighter lowered the fly section of the ladder. The wall began to collapse, and the firefighter under the ladder was struck and partially buried by falling bricks and debris. He was quickly removed from the hazard area, and firefighters began emergency medical treatment. He was transported by ambulance to a local emergency room and then flown by medical helicopter to a regional hospital where he later died. His death was caused by a head injury.

- Firefighters found a working fire upon their arrival at an apartment building and were ordered to vent the roof. The firefighters accessed the roof over a ground ladder and prepared to make their first cut. As one firefighter was walking toward the ridge of the roof, the structure failed, and he dropped into the attic. After several attempts to rescue him, the firefighter was removed from the structure and provided with emergency medical treatment. Despite all efforts of firefighters, the injured firefighter died as the result of smoke inhalation.

- Firefighters were dispatched to a report of a fire in a three-story wood-frame residence. Upon arrival, they found a working fire to the rear of the structure. Responding to reports of a civilian trapped on the third floor, firefighters conducted a primary search. After the search was completed, the IC ordered all firefighters from the building and implemented defensive firefighting operations. A resident of the building continued to insist that his roommate was still trapped in the building. The fire had been knocked down, and it was determined that another search of the structure would be conducted in hopes of locating the missing roommate. Approximately 4 minutes into the search, a structural collapse occurred and trapped firefighters. One of the firefighters was crushed and fatally injured. The other firefighter was rescued from the building but was seriously injured.

Struck by Object

Being struck by an object was the cause of four fatal firefighter injuries in 2011.

- Five fire departments were dispatched to a wildland fire. One of the firefighters drove a brush truck to the incident along with another firefighter and joined other units in attacking the eastern flank of the fire from the unburned area. Access to the area was provided by a single gate from a roadway. A wind shift drove the fire toward firefighters, forcing an immediate evacuation of the area. As apparatus left the area, firefighters on the first brush trucks to escape yelled to firefighters staged by the roadway to immediately evacuate. A brush truck stopped at the gate and was abandoned by the firefighters assigned to it. This blocked the escape path for the deceased and the other firefighter on his apparatus. Both firefighters abandoned their vehicle and ran to the roadway to escape the fire. The firefighters were separated during the escape. After the fire passed, one firefighter, the driver of the brush truck, was discovered deceased in a ditch. Initial assumptions were that he died due to fire exposure injuries. During the autopsy, it was discovered that he had died of blunt trauma. He was apparently run over by one of the fire vehicles leaving the scene as the fire approached and passed.

- One firefighter was killed when the jack that was supporting a fire department vehicle he was under failed. Firefighters used a hydraulic rescue tool to lift the vehicle off the trapped firefighter, and he was transported to the hospital by ambulance where he later died. The cause of death was listed as asphyxiation due to compression of the torso.

- One firefighter was killed while directing traffic around a vehicle crash. He was wearing a coat or jacket with some reflective material. A vehicle approaching the scene was going too fast, made a last-minute lane change, and struck the firefighter. He was pronounced dead at the scene.

- One firefighter wearing a reflective traffic vest was struck and killed by a passing vehicle while directing traffic at the scene of an MVA. The firefighter was transported to the hospital by medical helicopter but pronounced dead due to traumatic injuries shortly after her arrival.

Fall

Six firefighters died in 2011 as the result of falls in five separate incidents.

- One firefighter was responding to a carbon monoxide alarm from his residence. As he responded, he slipped on ice in his driveway and sustained a serious spinal injury. The firefighter was transported to the hospital but died later as the result of his injuries.

- One firefighter died from injuries sustained after rope rescue training at his fire station. Training had been completed when the firefighter attempted to climb one of two ropes suspended below the raised platform of the department's ladder tower. He likely lost his grip and fell 6 to 8 feet to the ground. He sustained a fatal head injury in the fall.

- A firefighter was attempting to remove a metal sign from an exterior wall at a fire department station when he fell from a 6-foot ladder and struck his head. He suffered a fatal injury and was pronounced dead at the scene.

- Fire department members were fighting a fire in a large coal bin in their community. The bin was approximately 75 feet tall. Firefighters had been working for some time to remove coal from the bin and believed that the incident was under control. About 80 tons of coal had already been removed with approximately 40 tons remaining in the bin. Two firefighters, cousins, were on the top of the bin flowing water into the bin to control the fire, when an explosion occurred. The explosion ripped off the roof and one wall of the bin. Both firefighters were trapped in the debris and killed. The State Fire Marshal's report determined that the explosion was caused by spontaneous combustion. News reports explained that the type of coal in use, Powder River Basin coal, was a factor in the blast.

- Firefighters responded to a report of a pile of more than 100 railroad ties on fire. Fog and smoke obscured visibility in the area. Crews could smell smoke in the area but could not find the fire. One firefighter was on a local bridge looking for the fire when he fell from the bridge to the embankment below. He was killed in the fall. The fire was determined to be intentionally set. The alleged arsonist was charged with murder for the death of the firefighter.

Out of Air

One firefighter died when he ran out of air while operating at the scene of a fire in 2011.

- Fire department units were dispatched to an automatic fire alarm activation reporting smoke on the fifth floor of a highrise medical office building. Arriving firefighters found a working fire in Suite 500 of the building. Additional fire department resources were requested. Two firefighters were conducting a search of the fifth floor. Fire department operations were complicated by locked doors, extreme heat, low visibility, and difficulties with the building's standpipe system. One firefighter called a Mayday when he and the other firefighter were unable to exit the hazardous area. An additional alarm was called, and a Rapid Intervention Crew (RIC) was deployed, but the firefighters could not immediately be located. Once located, the injured firefighter was removed from the building and provided with emergency medical treatment. Despite these efforts, he died as a result of smoke inhalation. The fire was determined to have been intentional, set by someone in Suite 500.

Other

Four firefighters died in 2011 of a cause that is not categorized above.

- While on duty, a firefighter responded to three emergency incidents during his shift, two of which were fires. The firefighter felt a pull in his back while working at one of the fire incidents. Early the next morning, while still on duty, the firefighter complained to his CO about worsening back pain. He was transported to the hospital by ambulance. Once at the hospital, he was diagnosed with a tear in his aorta. He underwent emergency surgery and subsequently died of his injury.

- One firefighter, a Hotshot crew member, became separated and lost from fellow firefighters after completing their assignment and returning by foot in rough backcountry terrain. His body was discovered the following day, but the cause and circumstances related to his death were not released.

- A firefighter responded to two emergency incidents. At the first incident, the firefighter was the driver of a brush truck that helped to extinguish a brush fire. The incident concluded at 1540 hours. The second incident was a report of smoke in a residence. The firefighter responded to the fire station where he stood by until the incident was concluded at 1922 hours. At approximately 1330 hours the following day, the firefighter was transported to the hospital for difficulty breathing and a rapid heart rate. He was assessed at the hospital and released later that day. In the morning of the following day, the firefighter experienced the same symptoms as the previous day. He went to the hospital and was pronounced dead a short time later. The cause of death was a pulmonary embolism.

- A firefighter, wearing full structural protective clothing and an SCBA, entered an apartment building to fight a fire and collapsed while working on the second floor. He was removed from the structure by other firefighters and treated at the scene. He was rushed to the hospital by ambulance but did not survive. The nature of the firefighter's fatal injuries has not been disclosed.

Nature of Fatal Injury

Figure 9 shows the distribution of the 83 firefighter deaths that occurred in 2011 by the medical nature of the fatal injury or illness. For heart attacks, Figure 10 shows the type of duty involved.

The National Institute for Occupational Safety and Health's (NIOSH's) Fire Fighter Fatality Investigation and Prevention Program (http://www.cdc.gov/niosh/fire/) is the leading resource in the world for investigations and focused reports on firefighter fatality incidents in the United States of America.

Figure 9. Fatalities by Nature of Fatal Injury (2011)

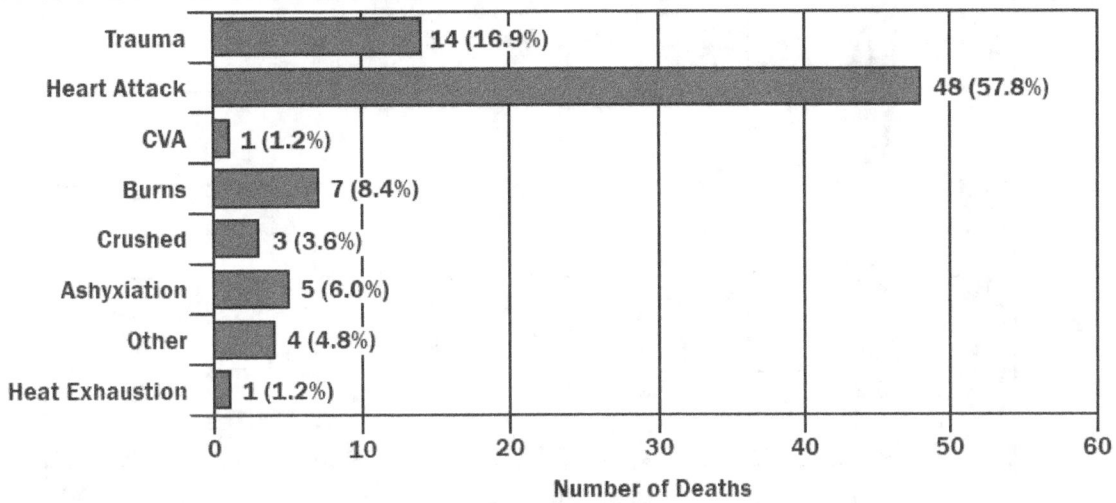

Figure 10. Heart Attacks by Type of Duty (2011)

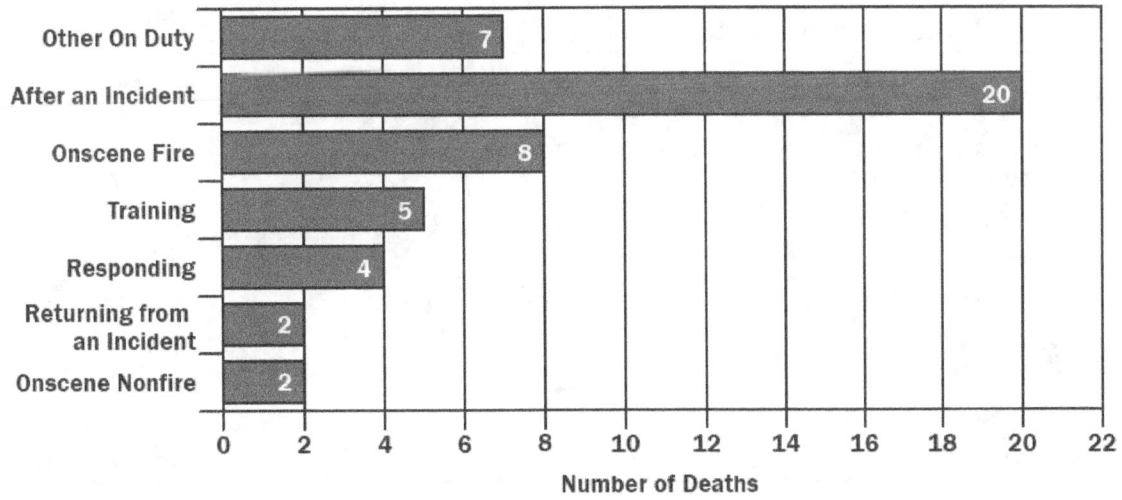

FIREFIGHTER AGES

Figure 11 shows the percentage distribution of fire-fighter deaths by age and nature of the fatal injury. Table 11 provides a count of firefighter fatalities by age and the nature of the fatal injury.

Younger firefighters were more likely to have died as a result of traumatic injuries, such as injuries from an apparatus accident or becoming caught or trapped during firefighting operations. Stress-related deaths are rare below the 31 to 35 years of age category and, when they occur, often include underlying medical conditions.

Figure 11. Fatalities by Age and Nature (2011)

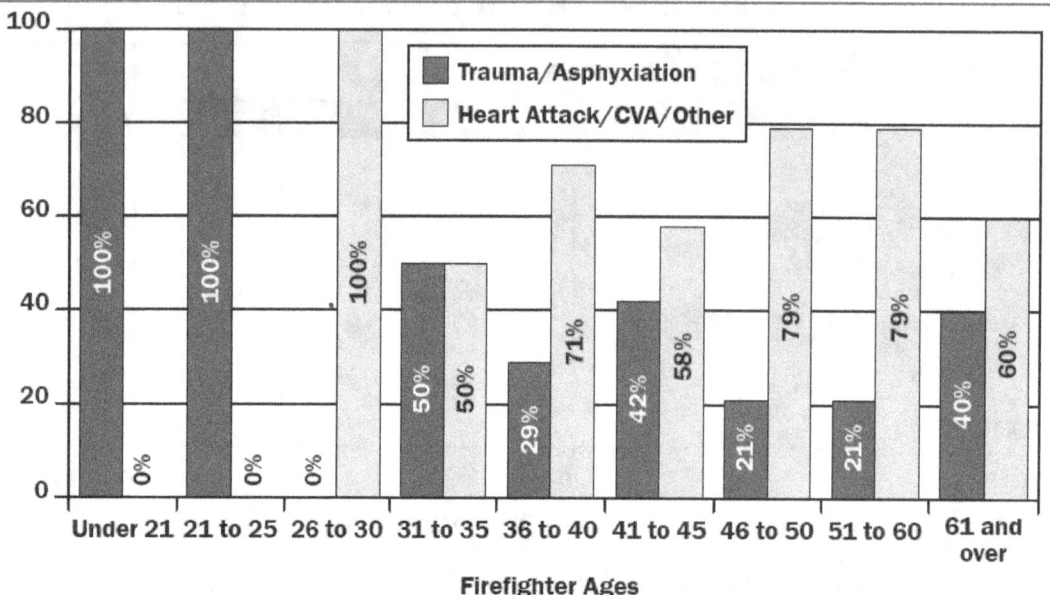

Firefighter Ages

Table 11. Firefighter Ages and Nature of Fatal Injury (2011)

Age Range	Heart Attack/CVA/Other	Trauma/Asphyxiation Total
under 21	0	2
21 to 25	0	7
26 to 30	2	0
31 to 35	3	3
36 to 40	5	2
41 to 45	7	5
46 to 50	11	3
51 to 60	22	6
61 and over	3	2

In 2011, the youngest firefighter and the only teenager to die while on duty was age 18. He died in a crash avoiding oncoming traffic in his lane while responding to an incident in his privately-owned vehicle (POV). The oldest firefighter killed on duty in 2011 was 82. He suffered a traumatic injury from a fall while responding to an incident.

Deaths by Time of Injury

The distribution of 2011 firefighter deaths according to the time of day when the fatal injury occurred is illustrated in Figure 12. The time of fatal injury for 4 firefighters was either unknown or not reported.

Figure 12. Fatalities by Time of Fatal Injury (2011)

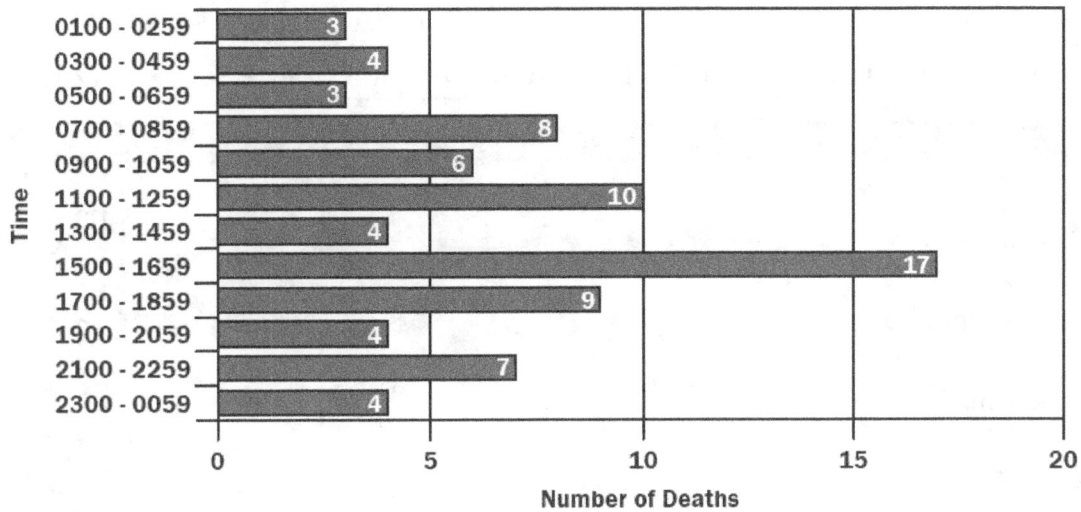

Firefighter Fatality Incidents by Month of Year

Figure 13 illustrates the 2011 firefighter fatalities by month of the year.

Figure 13. Deaths by Month of Year (2011)

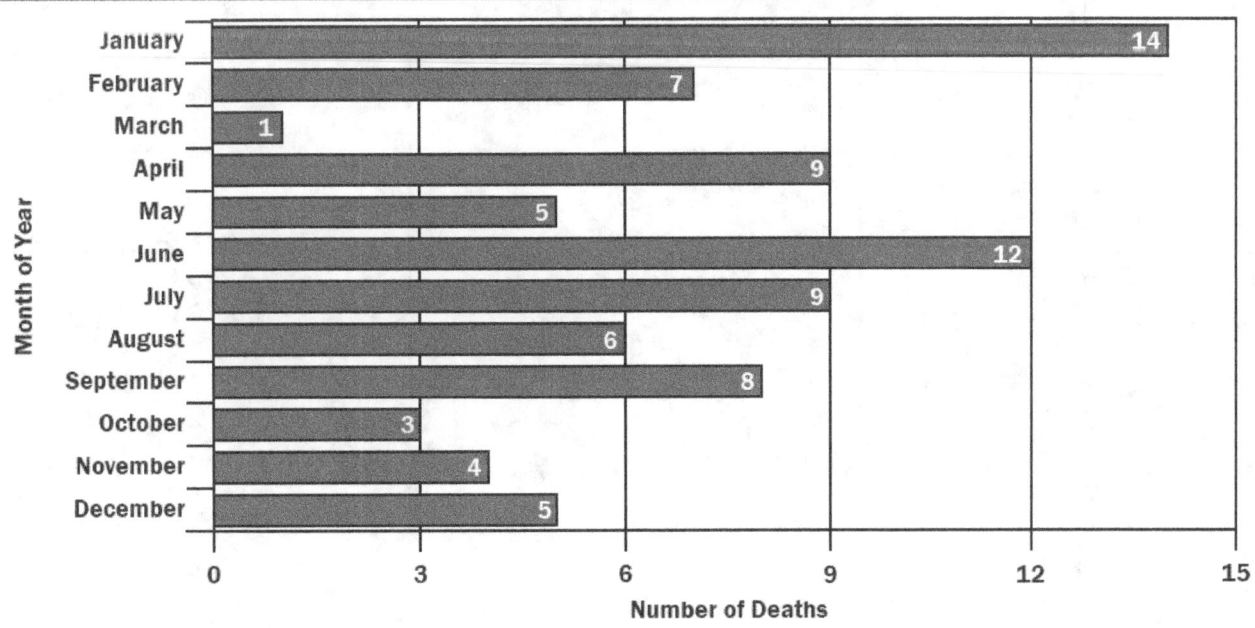

State and Region

The distribution of firefighter deaths in 2011 by State is shown in Table 12. Firefighters based in thirty-four States, plus one Air Force firefighter stationed overseas, died in 2011.

The highest number of firefighter deaths, based on the location of the fire service organization in 2011, occurred in North Carolina with seven deaths. Texas had the next highest total of firefighter fatalities in 2011 with six firefighter deaths.

Table 12. Firefighter Fatalities by State by Location of Fire Service* (2011)

State	Fatalities	Percentage
AR	1	1.2
AZ	1	1.2
CT	1	1.2
GA	1	1.2
GU	1	1.2
IA	1	1.2
KS	1	1.2
MD	1	1.2
ME	1	1.2
MN	1	1.2
NH	1	1.2
OK	1	1.2
WI	1	1.2
WV	1	1.2
MIL	1	1.2
SC	2	2.4
UT	2	2.4
VA	2	2.4
WA	2	2.4
LA	2	2.4
KY	3	3.6
MA	3	3.6
MO	3	3.6
NJ	3	3.6
PA	3	3.6
SD	3	3.6
IL	3	3.6
FL	4	4.8
IN	4	4.8
MS	4	4.8
NY	4	4.8
OH	4	4.8
CA	4	4.8
TX	6	7.2
NC	7	8.4

*This list attributes the deaths according to the State in which the fire department or unit is based, as opposed to the State in which the death occurred. They are listed by those States for statistical purposes and for the National Fallen Firefighters Memorial at the National Emergency Training Center (NETC). Due to rounding, percentage totals may not add up to 100.

Figure 14. Firefighter Fatalities by Region (2011)

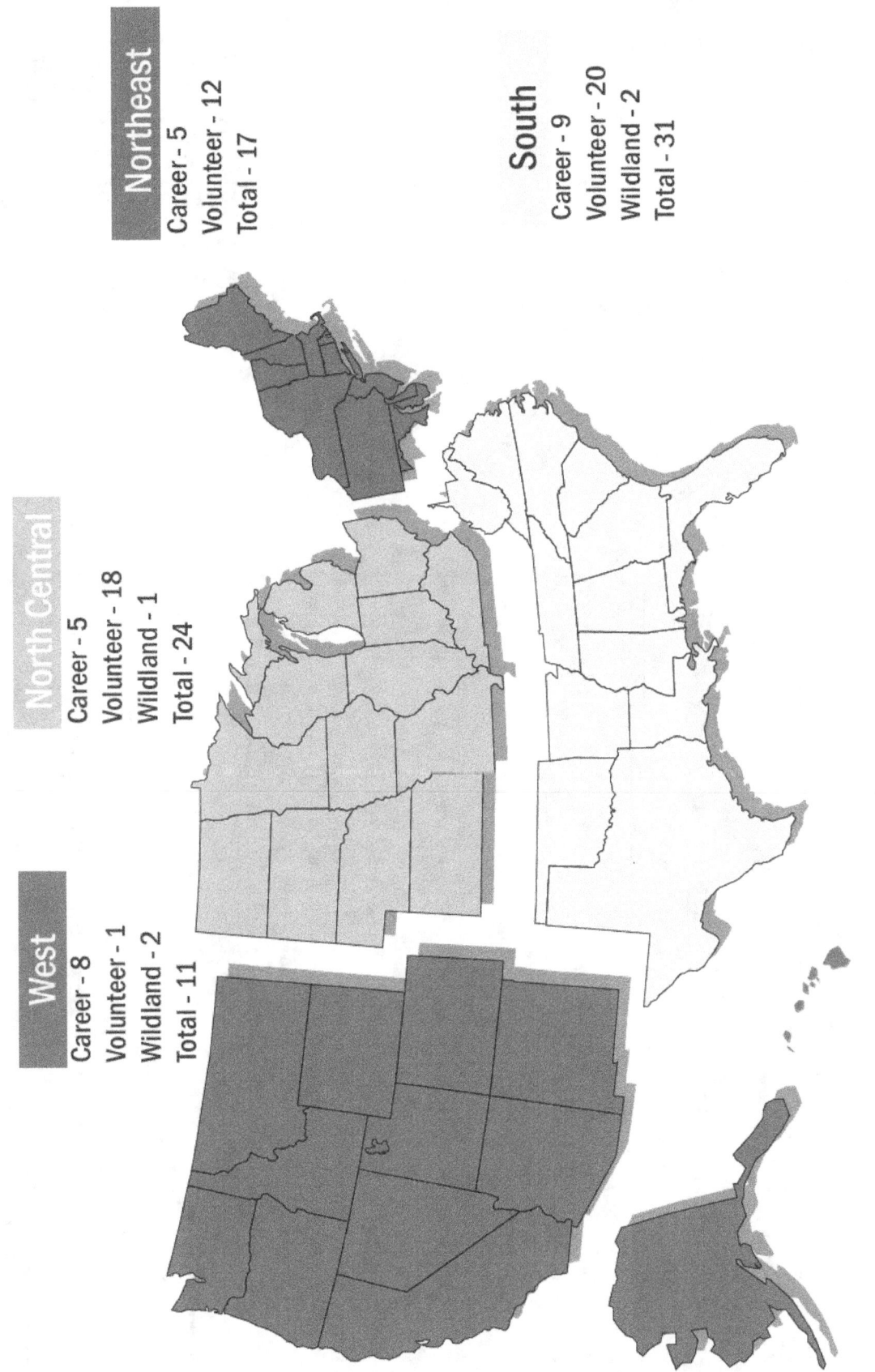

Northeast
Career - 5
Volunteer - 12
Total - 17

South
Career - 9
Volunteer - 20
Wildland - 2
Total - 31

North Central
Career - 5
Volunteer - 18
Wildland - 1
Total - 24

West
Career - 8
Volunteer - 1
Wildland - 2
Total - 11

Figure 15. Onduty Firefighter Fatalities by Fire Department Location (2011).

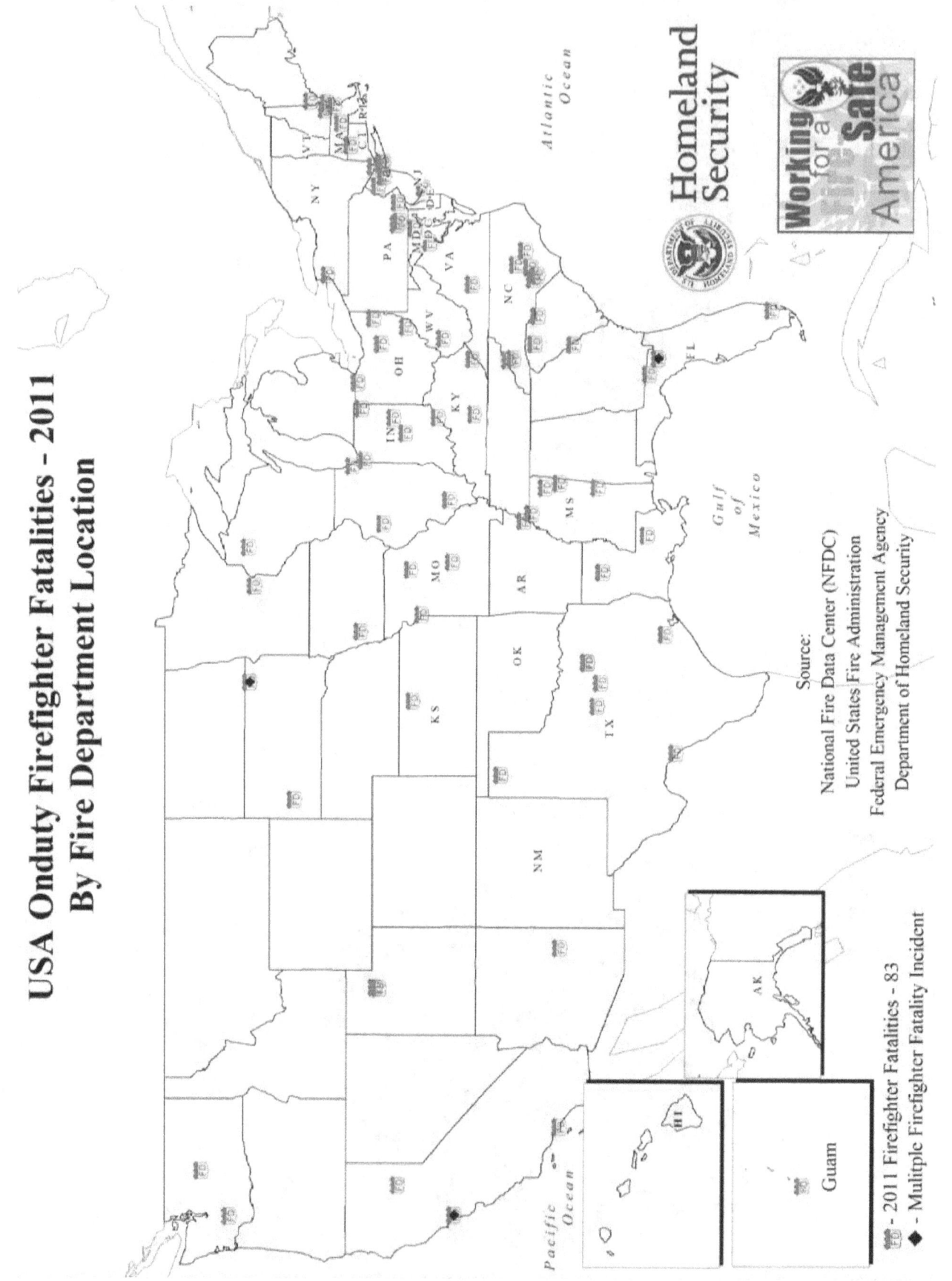

Figure 16. Onduty Firefighter Fatalities by Incident Location (2011)

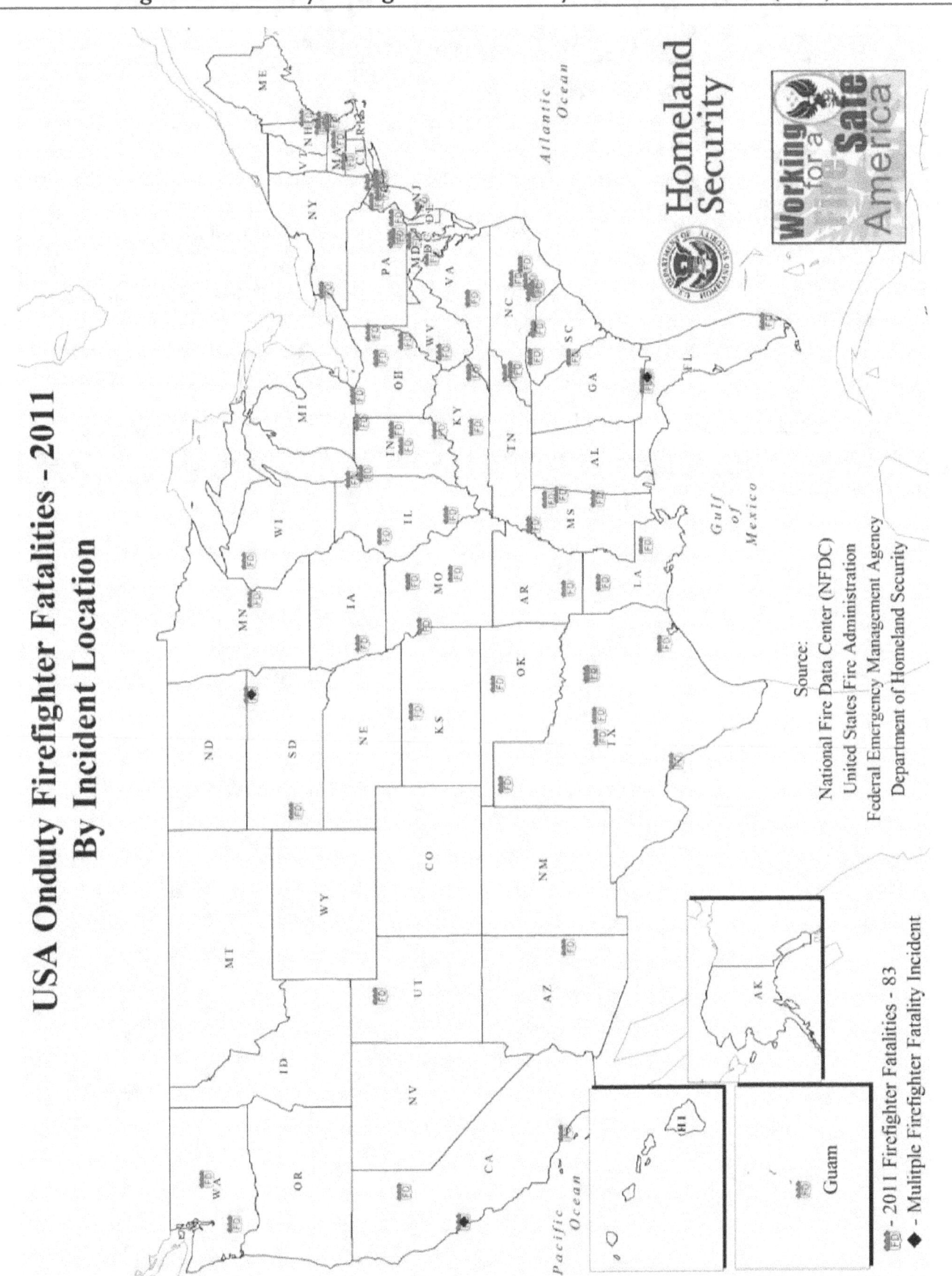

Analysis of Urban/Rural/Suburban Patterns in Firefighter Fatalities

The U.S. Census Bureau defines "urban" as a place having a population of at least 2,500 or lying within a designated urban area. "Rural" is defined as any community that is not urban. "Suburban" is not a census term but may be taken to refer to any place, urban or rural, that lies within a metropolitan area defined by the Census Bureau, but not within one of the central cities of that metropolitan area.

Fire department areas of responsibility do not always conform to the boundaries used by the Census Bureau. For example, fire departments organized by counties or special fire protection districts may have both urban and rural coverage areas. In such cases, where it may not be possible to characterize the entire coverage area of the fire department as rural or urban, firefighter deaths were listed as urban or rural based on the particular community or location in which the fatality occurred.

The following patterns were found for 2011 firefighter fatalities. These statistics are based on answers from the fire departments, and when no data from the departments were available, the data were based upon population and area served as reported by the fire departments.

Table 13. Firefighter Deaths by Coverage Area Type (2011)

	Urban/Suburban	Rural	Total
Firefighter Deaths	42	41	83

Appendix A

In memory of all firefighters
who answered their last call in 2011

To their families and friends

To their service and sacrifice

January 3, 2011–2159 hours
David Earl Remington, Sr., Firefighter
Age 58, Volunteer • Shapleigh Fire and Rescue Department, Maine

Firefighter Remington and the members of his fire department were dispatched to a motor vehicle crash at 2146 hours. Firefighter Remington notified dispatch by portable radio that he was responding. Firefighter Remington responded in his personal vehicle, a 2005 K1500 Chevrolet 4 wheel drive pickup truck.

As Firefighter Remington responded, the right wheels of his vehicle left the roadway in a gradual curve. Firefighter Remington attempted to bring the truck back onto the roadway by steering hard to the left. The truck turned sideways and slid diagonally across the roadway until it began to roll. Firefighter Remington was ejected during the roll and was found 15–20 feet away from his truck when it came to rest.

Firefighter Remington received fatal head injuries in the crash. Responding firefighters and emergency personnel pronounced him dead at the scene. Speed and the failure to wear a seatbelt were cited as factors in the crash.

January 7, 2011–1030 hours
Richard Eugene Paul, Fire Apparatus Operator
Age 54, Career • Kansas City Fire Department, Missouri

Fire Apparatus Operator Paul and the members of his company reported to a medical office to complete their annual physical examination. Fire Apparatus Operator Paul completed other elements of the physical examination and began the treadmill portion of the activity.

Approximately 3 minutes into the treadmill test, Fire Apparatus Operator Paul became ill. As treatment began, Fire Apparatus Operator Paul suffered a heart attack. Medical staff and other firefighters began cardiopulmonary resuscitation (CPR). Fire Apparatus Operator Paul was transported to the hospital but did not recover. The cause of death was listed as atherosclerotic cardiovascular disease.

For additional information regarding this incident, please refer to NIOSH Fire Fighter Fatality Investigation and Prevention Program report F2011-04 (www.cdc.gov/niosh/fire/reports/face201104.html).

January 10, 2011–Time Unknown
William Floyd (Bill) Hopman, Captain
Age 53, Volunteer • Quincy Fire Protection District, California

Captain Hopman responded as the driver of an engine company to a structure fire at 2108 hours on January 10, 2011. Upon arrival on the scene, Captain Hopman was assigned fire suppression duties on the second story of a residential structure with fire in the walls and attic. He wore full structural firefighting protective clothing, including a self-contained breathing apparatus (SCBA).

The fire was controlled and Captain Hopman drove the engine back to quarters and placed it back into service. Captain Hopman arrived home at approximately 2315 hours, showered, and went to bed without speaking to anyone. He was found deceased in his bed at 0719 hours the next morning by a family member. The cause of death was listed as a heart attack.

January 10, 2011–2200 hours
Jarrett Tyrell Eleam, Lieutenant
Age 26, Volunteer • Big Tree Volunteer Fire Company, Blasdell, New York

Lieutenant Eleam and the members of his fire department were participating in SCBA training. After a medical assessment, Lieutenant Eleam donned full structural firefighting protective clothing, including an SCBA. For approximately 16 minutes Lieutenant Eleam and other firefighters simulated firefighting tasks by crawling through an obstacle course.

After exiting the course, Lieutenant Eleam removed his SCBA and protective coat and proceeded to the rehabilitation section of the training exercise. Vital signs were taken, and Lieutenant Eleam drank some fluids while he recovered. He was released from rehabilitation and went back to the training activity where he observed and encouraged other firefighters.

After approximately 2 hours, the training activity had been completed. Lieutenant Eleam complained of feeling ill. He was assessed by firefighters and transported to the hospital. Upon his arrival at the hospital at 2306 hours, Lieutenant Eleam was treated and admitted. At approximately 0620 hours the next morning, Lieutenant Eleam's condition worsened. Testing eventually revealed a blood clot in his brain. Lieutenant Eleam died on January 12, 2011.

For additional information regarding this incident, please refer to NIOSH Fire Fighter Fatality Investigation and Prevention Program report F2011-26 (www.cdc.gov/niosh/fire/reports/face201126.html).

January 13, 2011–1224 hours
Jimmie Duane Niles, Captain
Age 59, Volunteer • Downs Fire Department, Kansas

Captain Niles and the members of his fire department responded to a garage fire. Captain Niles was at home at the time of dispatch and drove his personal vehicle to the fire station. Upon arrival at the station, he complained of shortness of breath. He donned his protective clothing and drove a brush firefighting apparatus to the scene.

Once at the scene, Captain Niles used an apparatus-mounted remote-controlled nozzle to assist with fire control. He remained in the cab of the apparatus while the other firefighter who had ridden in the cab dismounted and assisted with overhaul. A few minutes later, firefighters discovered Captain Niles slumped over the steering wheel in the apparatus cab.

Captain Niles was removed and CPR was initiated by firefighters and three local nurses who happened to be on the scene. He was transported to the hospital where he was later pronounced dead. The cause of death was listed as an enlarged heart and narrowing of the arteries.

January 15, 2011–Time Unknown
Patrick Bernard Hannon, Lieutenant
Age 51, Career • Chicago Fire Department, Illinois

Lieutenant Hannon arrived on duty on January 14, 2011 at 0630 hours. During the shift, Lieutenant Hannon's ladder company responded to four emergency incidents, including an emergency medical incident and two move-up assignments.

After lunch, Lieutenant Hannon supervised and participated in a training activity with the ladder's hydraulic extrication tools. Lieutenant Hannon handled the tools and instructed his firefighters on vehicle extrication techniques.

Lieutenant Hannon retired to his quarters at approximately 2230 hours. At approximately 0610 hours the next morning, firefighters noted that he had not risen at his usual time, so they knocked on the door of his bedroom. When no response was received, firefighters entered the bedroom and found Lieutenant Hannon deceased in bed. Despite efforts by firefighters and transportation by ambulance to the hospital, Lieutenant Hannon was not revived. The cause of death was listed as a heart attack.

January 16, 2011–1100 hours
Harold Frederick Frey, Firefighter/Paramedic
Age 46, Volunteer • Sandown Fire/Rescue, New Hampshire

Firefighter/Paramedic Frey was participating in ice water rescue training in the Exeter River. He wore an ice rescue protective suit and was in the water for approximately 30 minutes. After leaving the water under his own power, Firefighter/Paramedic Frey complained to other firefighters about shortness of breath and laid himself down in the snow. As firefighters carried Firefighter/Paramedic Frey to a rescue unit, Firefighter/Paramedic Frey began to have seizure-like activity.

Firefighters began CPR and advanced life support (ALS) emergency medical procedures. An ambulance arrived, and Firefighter/Paramedic Frey was transported to the hospital. His condition did not improve, and he was pronounced dead in the emergency room. The cause of death was a heart attack.

For additional information regarding this incident, please refer to NIOSH Fire Fighter Fatality Investigation and Prevention Program report F2011-03 (www.cdc.gov/niosh/fire/reports/face201103.html).

January 19, 2011–1816 hours
Mark Gray Falkenhan, Firefighter/Paramedic
Age 43, Volunteer • Baltimore County Fire Department–Lutherville Volunteer Fire Company, Maryland

Baltimore County firefighters were dispatched to a report of a kitchen fire in a three-story garden-style apartment building. Upon arrival, firefighters reported smoke showing and a civilian at a third floor window threatening to jump. Firefighters rescued the civilian over a ground ladder and stretched attack lines into the building. The Incident Commander (IC) requested a second alarm. Fire was found in multiple apartments, and a second unconscious civilian victim was discovered and removed by firefighters.

Firefighter/Paramedic Falkenhan, acting as the unit's Company Officer (CO), arrived on a heavy rescue truck approximately 11 minutes after the arrival of the first unit. Firefighter/Paramedic Falkenhan and another firefighter proceeded to the second floor of the fire occupancy. After searching the second floor and reporting high-heat conditions, they proceeded to the third floor to continue searching the structure.

The two firefighters completed the search of one third-floor apartment and began to search a second apartment. Rapid fire progress occurred and trapped Firefighter/Paramedic Falkenhan and the other firefighter in the apartment. The IC observed the advancement of the fire and ordered an evacuation. Tones were sounded over the radio, and air horns on fire apparatus were sounded. The firefighter with Firefighter/Paramedic Falkenhan was able to make it to a window and down a ladder.

Continued on next page.

Firefighter/Paramedic Falkenhan declared a Mayday and advised the IC that he was trapped on the third floor. Firefighters accessed the apartment where Firefighter/Paramedic Falkenhan was trapped over ground ladders and an aerial ladder. He was located and removed from the structure. Despite efforts by firefighters and medical personnel, Firefighter/Paramedic Falkenhan died of injuries he received in the fire. His death was caused by burns.

The Baltimore County Fire Department prepared a report on this incident. The report can be located at the Baltimore County website (http://resources.baltimorecountymd.gov/Documents/Fire/report/finalreport120322.pdf).

January 20, 2011–1605 hours
Leslie Leonard "Les" Clark, Captain
Age 80, Volunteer • Dixon Rural Fire Protection District, Missouri

Captain Clark and the members of his fire department were dispatched to a chimney fire in their community. Captain Clark was driving a tanker (tender) apparatus to the scene and reported to the tanker's passenger that he was feeling ill. The passenger activated the apparatus parking brake and stopped the vehicle.

Firefighters removed Captain Clark from the apparatus and began CPR. An ambulance arrived, and Captain Clark was transported to the hospital by medical helicopter. He did not recover. His death was caused by a heart attack.

January 26, 2011–Time Unknown
Thomas V. Regan, Firefighter
Age 82, Volunteer • Garden City Park Fire Department, New York

Firefighter Regan was responding to a carbon monoxide alarm from his residence. As he responded, he slipped on ice in his driveway and sustained a serious spinal injury. Firefighter Regan was transported to the hospital but died as the result of his injuries on February 17, 2011.

Firefighter Regan was a retired New York City Deputy Fire Chief. He was posthumously elected to the position of Chief of the Garden City Park Fire Department.

January 26, 2011–1530 hours
David Lewis Eason, Firefighter Recruit
Age 37, Career • West Memphis Fire Department, Arkansas

Firefighter Recruit Eason was enrolled in a resident firefighter certification program at the Arkansas Fire Training Academy. As a part of the training exercises scheduled for the day, Firefighter Recruit Eason was required to complete an SCBA confidence course in full structural firefighting protective clothing, including an SCBA. During the first training session of the day, Firefighter Recruit Eason had failed to complete this activity when he ran out of air.

At approximately 1515 hours, Firefighter Recruit Eason began the SCBA confidence exercise retest. As he crawled through the prop, his SCBA became hung up on a wooden board. Instructors observing the exercise provided assistance and freed Firefighter Recruit Eason's SCBA. Moments later, Firefighter Recruit Eason became unresponsive. Instructors removed him from the prop, CPR was initiated, and an Automated External Defibrillator (AED) was attached. Firefighter Recruit Eason was treated and transported by ambulance to the hospital where he was later pronounced dead. The cause of death was a heart attack.

For additional information regarding this incident, please refer to NIOSH Fire Fighter Fatality Investigation and Prevention Program report F2011-08 (www.cdc.gov/niosh/fire/reports/face201108.html).

January 29, 2011–1413 hours
Antonio Everett "Tony" Jones, Firefighter
Age 45, Career • Augusta Fire Department, Georgia

Firefighter Jones was on duty in his fire station when he began to feel ill. He requested and was granted emergency medical leave to see his doctor. He departed the station in his personal vehicle.

A short time after he left the station, firefighters were dispatched to a report of an unconscious man in a vehicle. Firefighters discovered Firefighter Jones in his vehicle and transported him to the hospital. He was later pronounced dead as the result of a heart attack.

January 29, 2011–1500 hours
Richard Leo "Ricky" Barbour, Fire Chief
Age 55, Volunteer • Wilson's Mills Fire Department, North Carolina

Chief Barbour and the members of his fire department responded to a woods fire on January 29, 2011. The following day, within 24 hours of the woods fire response, Chief Barbour began to experience chest pains and was transported to the hospital. He died just before midnight on February 4, 2011, from a heart attack.

January 29, 2011–2100 hours
James Franklin Walters, Sr., Assistant Chief
Age 57, Volunteer • Parkton Volunteer Fire Department, North Carolina

Assistant Chief Walters had been at his fire station for most of the day making the department's apparatus ready for an upcoming weekend fire school. Assistant Chief Walters also assisted other firefighters with apparatus and equipment familiarization. At approximately 2100 hours, Assistant Chief Walters was preparing to leave the fire station when he suffered a heart attack in the apparatus bay.

Firefighters, responding law enforcement, and EMS workers provided CPR, applied an AED, provided ALS-level emergency medical care, and transported Assistant Chief Walters to the hospital. Assistant Chief Walters was pronounced dead due to a heart attack at 2214 hours.

February 1, 2011–Time Unknown
Steven Frederick Auch, Battalion Chief
Age 56, Career • Indianapolis Fire Department, Indiana

Battalion Chief Auch worked a shift from 0800 hours on January 31, 2011, through 0800 hours on February 1, 2011. During the shift, Battalion Chief Auch responded to two emergency incidents, including a car crash into a house and a reported structure fire. Battalion Chief Auch completed his shift and went home. He was observed by a neighbor at approximately 1230 hours.

At the request of Battalion Chief Auch's spouse, a neighbor went to the house at approximately 1530 hours to check on his status. His spouse had been unable to contact him and was concerned. The neighbor found Battalion Chief Auch deceased on the kitchen floor. The cause of death was listed as a heart attack.

February 1, 2011–2030 hours
Daniel Charles "Double Dare" Dare, Firefighter/First Responder
Age 52, Volunteer • Avon Fire Protection District, Illinois

Firefighter Dare and the members of his fire department were dispatched to a medical emergency involving a child. Firefighter Dare responded in his personal vehicle. The area was under blizzard conditions at the time of the call.

Firefighter Dare, who lived in a rural area, became stuck in the snow. Other firefighters also became stuck in the snow, and the emergency medical incident was eventually cancelled. A firefighter with a tractor pulled Firefighter Dare's vehicle out, and he drove home. Upon his arrival home, he suffered a heart attack while still in his vehicle.

Firefighter Dare's wife attempted to drive him to the fire station for treatment but also became stuck in the snow. The firefighter operating the tractor came and pulled Firefighter Dare's vehicle from the snow and towed it to the fire station.

Emergency medical treatment was provided by firefighters at the station, but no other emergency resources could reach the station due to the storm. Medical control authorized the termination of CPR, and Firefighter Dare was pronounced dead as the result of a heart attack.

February 11, 2011–1800 hours
Derek Kozorosky, Airman First Class
Age 22, Career • 18th Civil Engineer Squadron, United States Air Force, Kadena Air Base, Japan

Airman First Class Kozorosky and two other firefighters had completed a driver training exercise for a military P-23 Aircraft Rescue Fire Fighting (ARFF) apparatus. After completing service on the apparatus, including refilling of the water tank, Airman First Class Kozorosky directed the driver trainee to back the ARFF vehicle into an apparatus bay.

Airman First Class Kozorosky acted as the spotter near the rear of the apparatus as it backed up. The vehicle was misaligned in the bay, and, somehow, Airman First Class Kozorosky became trapped between a concrete column and the left rear tire of the ARFF vehicle. Airman First Class Kozorosky's feet were in front of the column and his torso and head were at the rear of the column.

After approximately 30 minutes of rescue efforts, the apparatus was moved, and Airman First Class Kozorosky was freed. He was transported to the hospital but died as a result of traumatic injuries.

An Air Force analysis of the incident concluded that Airman First Class Kozorosky should have been at the rear of the vehicle and that an additional certified driver and a spotter should have been used.

February 13, 2011–1600 hours
Joshua Jay "Josh" Wilkes, Firefighter
Age 26, Volunteer • Unity Fire Department, Mississippi

Firefighter Wilkes and the members of his fire department responded to a report of a fire involving the trailer of an 18-wheeler truck. The trailer was being used for storage, and fire was found coming from the roof as firefighters arrived.

Firefighter Wilkes and other firefighters extinguished the fire and remained on scene for approximately an hour completing overhaul. While en route back to the fire station, Firefighter Wilkes and another firefighter stopped at a convenience store to fill the apparatus with fuel. After entering the store, Firefighter Wilkes suffered a heart attack. He was treated at the scene and at the hospital but did not survive.

February 16, 2011–0002 hours
Glenn Leroy Allen, Firefighter/Paramedic
Age 61, Career • Los Angeles City Fire Department, California

At 2320 hours on February 15, 2011, Firefighter/Paramedic Allen and the members of his engine company were dispatched as a part of a larger response assignment to a structure fire in a sizeable (12,500 square foot) residence.

Firefighters found fire and smoke showing upon their arrival. Further investigation revealed fire in the walls of the structure that had spread to a large void ceiling space. Firefighters advanced hoselines into the structure, opened the roof, and began to open walls and ceilings on the interior.

Firefighter/Paramedic Allen's engine arrived approximately 14 minutes into the incident. The crew was ordered to assist in the interior and went inside with a Thermal Imaging Camera (TIC), pike poles, and an additional handline. The interior ceilings were high and longer pike poles and ground ladders were used. Fire appeared to be running through the attic, and firefighters struggled to expose all areas of fire.

At 0002 hours, a significant portion of the drop ceiling on the first floor of the structure fell onto interior firefighters. Three firefighters, including Firefighter/Paramedic Allen, were trapped in or under the debris. The IC requested additional resources, and rescue efforts were begun. After extensive efforts, Firefighter/Paramedic Allen was removed from the building at 0016 hours. He was transported by ambulance to the hospital but died as a result of his injuries on February 18, 2011.

The cause of the ceiling collapse was due to the failure of piping in the void space. Water accumulation and absorption by the ceiling caused it to fall and crush the firefighters working below.

February 17, 2011–1730 hours
Larry Cleveland Gressett, Sr., Firefighter
Age 33, Volunteer • Toomsuba/Alamucha Volunteer Fire Department, Mississippi

Firefighter Gressett and other firefighters responded to a local pond for a report of a person in the water in distress. When they arrived on the scene, firefighters observed a boat approximately 30 feet from the shore with one person in the boat supporting an unconscious person in the water.

A firefighter entered the water and began to swim out to the boaters in distress but could not continue due to the cold. Firefighter Gressett entered the water and swam for approximately 20 feet until he became incapacitated. Firefighter Gressett called for help, disappeared underwater, briefly reappeared, and went under water again.

A civilian entered the water and removed Firefighter Gressett. He was treated in an ambulance while en route to the hospital and also treated at the emergency room. He was pronounced dead at the hospital due to drowning. The boater who had been in distress also died.

February 24, 2011–Time Unknown
Christopher Tomhave Stock, Fire Chief
Age 48, Volunteer • Westport Volunteer Fire Department, Kentucky

On February 23, 2011, Chief Stock conducted wildland firefighting training for the members of his fire department. The training included water supply, pump operations, apparatus placement information, and practical exercises. The training was conducted from 1900 hours to 2200 hours. At approximately 1824 hours on February 24, 2011, Chief Stock was discovered unconscious by a family member.

Chief Stock was transported to a local hospital but later pronounced dead. His death was caused by a heart attack.

March 12, 2011–1912 hours
James Lee von Roden, Captain
Age 49, Volunteer • Lee Community Volunteer Fire Department, Florida

Captain von Roden and the members of his fire department were dispatched to a wildland fire incident in their community. Captain von Roden had responded to five incidents earlier in the day.

Between 30 and 60 minutes after he left his home to respond, Captain von Roden was found by his spouse in the front yard near his pickup truck, phone in hand. Fire and medical responders came to the scene to provide care, but Captain von Roden had already passed away. The cause of death was listed as a heart attack.

April 7, 2011–1745 hours
David Jerome "Davey" Hunsinger, Jr., Firefighter
Age 23, Volunteer • Tar Heel Rural Volunteer Fire Department, North Carolina

Firefighter Hunsinger was responding in his personal vehicle to a report of a brush fire. During the response, the right wheels of Firefighter Hunsinger's Mazda 3 left the right side of the roadway. In response, he overcorrected and crashed into a minivan headed in the opposite direction. Firefighter Hunsinger was pronounced dead at the scene of the crash.

News reports quoting the law enforcement report on the crash cited speed as a factor.

April 9, 2011–1630 hours
Elias Jaquez, Firefighter
Age 49, Volunteer • Cactus Volunteer Fire Department, Texas

Firefighter Jaquez and the members of his fire department were dispatched to a mutual-aid wildland fire near the Town of Dumas in the Texas Panhandle. Firefighter Jaquez and another firefighter responded to the incident in a brush truck. Firefighter Jaquez was assigned to run the pump and nozzle.

The Cactus Fire Department brush truck was paired with a brush truck from the Dumas Fire Department. The two units attacked the east flank of the fire from the unburned area. The Dumas brush truck became stuck in deep sand. The Cactus brush truck pulled alongside to allow the Dumas firefighters to come aboard. As the Cactus brush truck pulled away, it too became stuck in the sand.

Continued on next page.

As the fire advanced toward the two incapacitated trucks, the firefighters abandoned their positions and tried to outrun the fire. After the fire burned over the area, Firefighter Jaquez was found suffering from severe burns. He was transported to the hospital by ambulance but died as a result of his burns on April 20, 2011.

The fire eventually burned 35,000 acres. The Texas State Fire Marshal's Office prepared a detailed report on this incident. The report is available at http://www.tdi.state.tx.us/fire/fmloddinvesti.html

April 11, 2011–Time Unknown
Randy Dale Boley, Captain
Age 51, Volunteer • Clinton Township Volunteer Fire Department, Ohio

Captain Boley responded to two emergency medical incidents on April 11, 2011. The first required him to carry heavy equipment and perform CPR compressions for an extended period of time. The second incident was a lift assistance call to help an obese customer get back into a wheelchair. The second incident concluded at approximately 1023 hours. The balance of the day passed without incident, and Captain Boley went to bed at approximately 2200 hours.

The next morning at approximately 0600 hours, Captain Boley was discovered deceased on the floor. He had been dead for some time, so resuscitation was not attempted. The exact time of his death could not be determined.

April 13, 2011–0518 hours
Gregory Leon Harris, Firefighter
Age 40, Career • Miami Dade Fire Rescue Department, Florida

Firefighter Harris was on duty working on Aerial 38. During his shift, he responded to three emergency incidents during the shift, two of which were fires. Firefighter Harris felt a pull in his back while working at one of the fire incidents.

At 0518 hours the next morning, while still on duty, Firefighter Harris complained to his CO about worsening back pain. He was transported to the hospital by ambulance. Once at the hospital, he was diagnosed with a tear in his aorta. He underwent emergency surgery and subsequently died of his injury on April 22, 2011.

April 15, 2011–1300 hours
Gregory Mack Simmons, Firefighter
Age 50, Volunteer • Eastland Volunteer Fire Department, Texas

Five fire departments, including the Eastland Volunteer Fire Department, were dispatched to a wildland fire located to the south of the city limits of Eastland. Firefighter Simmons drove a brush truck to the incident along with another firefighter.

The Eastland brush truck joined other units and attacked the eastern flank of the fire from the unburned area. Access to the area was provided by a single gate from a roadway. A wind shift drove the fire toward the firefighters, forcing an immediate evacuation of the area.

As apparatus left the area, firefighters on the first brush trucks to escape yelled to firefighters staged by the roadway to evacuate immediately. A brush truck stopped at the gate and was abandoned by the firefighters assigned to it. This blocked the escape path for Firefighter Simmons and the other firefighter on his apparatus. Both firefighters abandoned their vehicles and ran to the roadway to escape the fire. The firefighters were separated during the escape.

After the fire passed, Firefighter Simmons was discovered deceased in a ditch. Initial assumptions were that he died due to fire exposure injuries. During the autopsy, it was discovered that Firefighter Simmons died of blunt trauma. He was apparently run over by one of the fire vehicles leaving the scene as the fire approached and passed.

The Texas State Fire Marshal's Office prepared a detailed report on this incident. The report is available at http://www.tdi.state.tx.us/fire/fmloddinvesti.html

For additional information regarding this incident, please refer to NIOSH Fire Fighter Fatality Investigation and Prevention Program report F2011-09 (www.cdc.gov/niosh/fire/reports/face201109.html).

April 17, 2011–1155 hours
Jacob Anthony Carter, Firefighter
Age 18, Volunteer • Becker-Athens Volunteer Fire Department, Mississippi

Firefighter Carter was at home when he and the members of his fire department were dispatched to a mutual-aid structure fire. Firefighter Carter responded in his personal vehicle, a 2002 Ford Explorer. His mother rode in the front passenger seat.

As he approached a curve, Firefighter Carter encountered a vehicle coming in the opposite direction. His mother stated that the oncoming vehicle was on their side of the road. Firefighter Carter steered to the right to avoid a collision. The right tires of his vehicle left the roadway, he overcorrected, and left the right side of the roadway. After leaving the roadway, his vehicle struck a tree, spun around and struck a second tree, then ended up back in the roadway.

Firefighter Carter was transported to the hospital by medical helicopter but did not survive his injuries. Firefighter Carter was wearing his seatbelt at the time of the crash. His mother received non-life threatening injuries.

April 18, 2011–2000 hours
Robert Dean Watts, Firefighter
Age 50, Volunteer • Windsor Volunteer Fire Department, Connecticut

Firefighter Watts rode in an engine company responding to a report of smoke at a local convalescent and rehabilitation facility. First-arriving command officers reported smoke from the structure. Prior to entering the scene, Firefighter Watts dismounted the engine and pulled a supply line to a fire hydrant.

Once the supply line had been laid, Firefighter Watts attached it to the fire hydrant and opened the hydrant. He removed a large kink in the 5-inch supply line by partially lifting and repositioning the hose. After completing the hydrant hookup, Firefighter Watts complained of not feeling well to paramedics on the scene and suddenly collapsed.

Firefighter Watts immediately received ALS emergency medical care and was transported by ambulance to the hospital. Firefighter Watts' condition never improved, and he was pronounced dead due to a heart attack.

April 25, 2011–2330 hours
Charles Edward Foster, Firefighter
Age 59, Volunteer • Barton Volunteer Fire Department, Mississippi

Firefighter Foster and the members of his department were clearing trees from a local roadway after a severe storm. As he worked on the second tree, Firefighter Foster suffered a heart attack. Other firefighters immediately began to provide treatment, and he was transported by ambulance to the hospital. Firefighter Foster did not recover and died on April 29, 2011.

April 26, 2011–2130 hours
Michael Conley Webb, Captain
Age 46, Volunteer • Fleming-Neon Volunteer Fire and Rescue Department, Kentucky

Captain Webb responded to the scene of a crash involving an ATV. Hours after returning from the incident, Captain Webb suffered a heart attack at his home and died.

May 18, 2011–0337 hours
Michael P. Esposito, Ex-Captain
Age 43, Volunteer • Baldwin Fire Department, New York

Ex-Captain Esposito and the members of his fire department were dispatched to a structure fire. While en route to the scene on an engine company, Ex-Captain Esposito complained of chest pains. When firefighters arrived at the scene, Ex-Captain Esposito was loaded into an ambulance and transported to the hospital.

At the hospital, Ex-Captain Esposito's condition worsened, and he died as a result of a heart attack.

May 22, 2011–Time Unknown
David Samuel Howell, Firefighter
Age 54, Volunteer • Roseboro Volunteer Fire Department, North Carolina

Firefighter Howell responded to a vehicle crash at approximately 1525 hours. Firefighter Howell responded directly to the scene, donned a traffic safety vest, and assessed the damage. Firefighter Howell, who reported to other firefighters that he was not feeling well, left the scene and went home. He stopped by the fire station later that evening and said that he was feeling better.

The next day, after not hearing from Firefighter Howell, firefighters went to his home to check his status and found him dead in his bed. Although no autopsy was performed, his death is presumed to be cardiac-related.

May 23, 2011–2130 hours
Chip Andrew Imker, Firefighter
Age 35, Volunteer • Cambridge Fire Department, Minnesota

Firefighter Imker participated in rope rescue training at his fire station. Training had been completed, and Firefighter Imker attempted to climb one of two ropes suspended below the raised platform of the department's ladder tower. He likely lost his grip and fell 6 to 8 feet to the ground. He sustained a head injury in the fall. Firefighter Imker was transported to the hospital where he died as a result of his injuries.

For additional information regarding this incident, please refer to NIOSH Fire Fighter Fatality Investigation and Prevention Program report F2011-12 (www.cdc.gov/niosh/fire/reports/face201112.html).

May 28, 2011–0756 hours
Robert J. Tieche, Fire Chief
Age 63, Career • Cardinal Joint Fire District, Canfield, Ohio

Chief Tieche worked a full day on May 27, 2011, including responses to two emergency incidents. He went home at 1600 hours after other firefighters observed that he did not look well. His wife said that he looked pale at home that evening.

At 0002 hours on May 28, 2011, Chief Tieche responded to an incident and supervised tanker refill operations. He cleared from the incident at 0121 hours.

At 0756 hours, firefighters responded to Chief Tieche's home. He was dizzy and appeared pale, with a low blood pressure. Chief Tieche was transported to the hospital by private vehicle. He was admitted to the hospital and died later that day as the result of a heart attack.

May 31, 2011–1230 hours
Thomas Michael Shields, Second Assistant Chief
Age 42, Volunteer • Flanders Fire & Rescue Company, New Jersey

Chief Shields and the members of his fire department were dispatched to a report of an ill child at a local school. The child was treated and transported to the hospital. Chief Shields cleared the incident at 1201 hours.

Chief Shields went to his fire station to complete paperwork and an inventory of radios. While at the fire station, he complained to another firefighter that he was hot and that he was experiencing chest pain. Chief Shields left the fire station at approximately 1230 hours and went home. Upon his arrival at home, he appeared pale and continued to complain about chest pain. Chief Shields' spouse, a registered nurse, began treatment for a heart attack and called for assistance.

Chief Shields was treated and transported to the hospital by ambulance. Chief Shields died later that day while undergoing medical care for a heart attack.

June 2, 2011–1045 hours
Vincent A. Perez, Lieutenant

Age 48, Career • San Francisco Fire Department, California

Anthony Michael "Tony" Valerio, Firefighter/Paramedic

Age 53, Career • San Francisco Fire Department, California

Lieutenant Perez and Firefighter/Paramedic Valerio were assigned to Engine 26. Their unit was dispatched as part of a larger assignment to a report of a structure fire in a residence. Engine 26 was the first unit on the scene and reported light smoke showing from the garage. Lieutenant Perez and Firefighter/Paramedic Valerio entered the structure to assess conditions. The firefighters stretched a 200-foot preconnected attack line into the structure. Both firefighters wore full structural firefighting protective clothing and SCBA.

As firefighters worked on the scene, a window in the fire room failed, allowing the introduction of additional oxygen to the fire. The fire progressed rapidly up a stairwell and overcame Lieutenant Perez and Firefighter/Paramedic Valerio. Firefighters could not advance hoselines into the building until the original fire was controlled.

Firefighters found and removed Lieutenant Perez and Firefighter/Paramedic Valerio from the building and initiated treatment. Despite their efforts, Lieutenant Perez died that day and Firefighter/Paramedic Valerio died on June 4, 2011. Both firefighters died as the result of burns.

The San Francisco Fire Department completed a comprehensive analysis and report of this incident. The report can be accessed at http://www.sf-fire.org/modules/showdocument.aspx?documentid=2694

June 10, 2011–1200 hours
Ronald Dwane Ruprecht, Lieutenant

Age 51, Volunteer • Stone Lake Fire Department, Wisconsin

Around noon on June 10, 2011, Lieutenant Ruprecht and the members of his fire department responded to an emergency. The incident was found to be a false alarm, and firefighters went back into service. Lieutenant Ruprecht was found dead at home by his parents the morning of June 11, 2011, after he failed to attend a family function. The autopsy indicated that the time of death was late on June 10 or in the early hours of June 11. The cause of death was cardiac-related.

June 12, 2011–0155 hours
Garet Gardner Rasmussen, Acting Battalion Chief
Age 38, Career • Chelan County Fire District 1, Washington

On June 1, 2011, Chief Rasmussen was promoted to Battalion Chief as a temporary assignment. As part of the requirements of that position, he worked "Command Duty." Command Duty requires the duty officer to remain on duty and available to respond to emergencies for a 1-week period that starts at 0800 hours on Wednesday morning, until the following Wednesday at 0800 hours. During the week, the duty officer will respond to emergencies from his or her home after normal business hours. At the time of his death, Chief Rasmussen was starting the fifth day of his very first week of Command Duty.

At about 0200 hours on Sunday morning, June 12, 2011, Battalion Chief Rasmussen was dispatched to an automobile crash with injuries. He was asleep at the time of the dispatch, and the response distance was far from his residence. The vehicle crash scene could not be located, and firefighters, including Chief Rasmussen, spent approximately 5–10 minutes searching the area. Despite the efforts of responders, the crash was never located, and firefighters returned to quarters.

Chief Rasmussen signed off the radio and returned to his home at around 0230 hours. Some time shortly after arriving home, Chief Rasmussen sat down in a chair. He was discovered in the chair, unresponsive, by his spouse at approximately 0730 hours. Battalion Chief Rasmussen died as the result of a heart attack.

June 15, 2011–1610 hours
Scott Thomas Davis, Firefighter
Age 40, Career • Muncie Fire Department, Indiana

Firefighter Davis and the members of his fire department were dispatched to a report of a fire in a large church. First-arriving firefighters reported visible flames and heavy smoke coming from the roof. The first-arriving CO called for a second alarm, and shortly thereafter called for additional tankers (tenders).

Firefighters entered the structure and found mostly clear conditions in the interior of the church. As firefighters began to open up the ceilings to access the attic space, they discovered a considerable amount of fire. Interior crews were having difficulty controlling the fire with a handline. Water supply was an issue; the area of the church did not have fire hydrants.

Interior firefighters notified the IC that they were withdrawing from the structure. As firefighters were preparing to leave, a structural collapse occurred. An accountability check was conducted, and it was realized that Firefighter Davis was missing.

Due to the volume of fire, firefighters were unable to access the collapsed area. Firefighter Davis was located when a news helicopter flying over the scene spotted his remains in the debris. The cause of death was listed as smoke inhalation. The origin of the fire was likely a lightning strike earlier in the day.

For additional information regarding this incident, please refer to NIOSH Fire Fighter Fatality Investigation and Prevention Program report F2011-14 (www.cdc.gov/niosh/fire/reports/face201114.html).

June 17, 2011–1530 hours
Corey Ray Shaw, Firefighter
Age 22, Paid-on-Call • Du Quoin Fire Department, Illinois

Firefighter Shaw and members of his fire department were dispatched to respond with their ladder truck to a working fire in a 96-year-old brick and masonry structure that housed an antique store and living quarters. Firefighter Shaw's department operates the only ladder apparatus in the county. Firefighter Shaw responded to the scene in his personal vehicle.

The fire fight was in a defensive mode with no interior firefighting operations. The fire had self-vented and cracks were visible in at least one exterior wall. It was noted that an extension ladder was against the building in an area that would be subject to collapse. Firefighter Shaw and another firefighter entered the collapse zone to retrieve the ladder. Firefighter Shaw took a position under the ladder, supporting the ladder as the other firefighter lowered the fly section of the ladder.

The wall began to collapse, and Firefighter Shaw attempted to flee. He was struck and partially buried by falling bricks and debris. Firefighter Shaw was quickly removed from the hazard area, and firefighters began emergency medical treatment. He was transported by ambulance to a local emergency room and then flown by medical helicopter to a regional hospital where he later died. His death was caused by a head injury.

For additional information regarding this incident, please refer to NIOSH Fire Fighter Fatality Investigation and Prevention Program report F2011-15 (www.cdc.gov/niosh/fire/reports/face201115.html).

June 18, 2011–Time Unknown
Robin Erlic West, Assistant Chief
Age 55, Volunteer • Startex Fire Department, Wellford, South Carolina

Assistant Chief West responded to two emergency incidents on June 18, 2011. The first was a structure fire where he was first on scene, and the second was a report of wires down due to a passing storm. The second incident concluded at approximately 1620 hours. Assistant Chief West returned to his home.

Later in the evening, Assistant Chief West suffered a medical emergency at his home. He was transported to the hospital and held overnight. He was released at approximately 1300 hours on June 19, 2011, with no diagnosis. He became ill again at 1545 hours and was transported back to the hospital. He did not recover and was pronounced dead of a heart attack at 1719 hours.

June 20, 2011–1610 hours
Brett Luther Fulton, Forest Ranger
Age 52, Wildland Full-Time • Suwannee Forestry Center, Florida Division of Forestry, Florida

Joshua Omer Burch, Forest Ranger
Age 31, Wildland Full-Time • Suwannee Forestry Center, Florida Division of Forestry, Florida

Forest Rangers Burch and Fulton were assigned to the Blue Ribbon Fire, a lightning-caused wildfire in Hamilton County, Florida. The fire had initially been contained on June 16, 2011, but it had spread beyond the containment area when checked on June 20.

Both Forest Rangers operated tractor/plow units. During operations, one of the units became stuck. The second unit and operator attempted to provide assistance. As the wildfire approached, both operators abandoned their units and attempted to outrun the fire. They were overtaken by the progress of the fire and died. Neither firefighter attempted to deploy a fire shelter. Both firefighters died of burns.

June 23, 2011–0620 hours
Chris Khuong Pham, Fire Rescue Officer
Age 35, Career • Dallas Fire-Rescue Department, Texas

Fire Rescue Officer Pham worked a 24-hour shift beginning on the morning of June 22, 2011. He was assigned to Rescue 29. During the course of the shift, he responded to 10 emergency medical incidents. The last of these incidents concluded at 0608 hours on June 23, 2011. Fire Rescue Officer Pham was last seen by crew members at approximately 0620 hours.

At approximately 0740 hours, Fire Rescue Officer Pham was discovered by other firefighters unresponsive in his bed. CPR was initiated and ALS procedures were given. He was transported by ambulance to the hospital but pronounced dead shortly after his arrival at the hospital. Fire Rescue Officer Pham's death was caused by a heart attack.

June 27, 2011–2211 hours
Matthew Morgan "Butch" Hadaller, III, Fire Chief
Age 47, Career • Lewis County Fire District #3, Washington

Chief Hadaller responded to two emergency medical incidents on the morning of June 27, 2011. The second incident concluded at approximately 0904 hours. Chief Hadaller spent the rest of the day completing vehicle inspection, maintenance, administrative, station, and physical fitness duties. He went home on standby at 2000 hours.

At 2211 hours, emergency medical assistance was sent to Chief Hadaller's home for a report of someone having a seizure. Chief Hadaller was found unconscious on the floor. He was treated and transported to the hospital but was pronounced dead at 2350 hours. The cause of death was cardiac-related.

June 30, 2011–0343 hours
Charles Victor "Sparky" Sparks, Firefighter
Age 49, Volunteer • Columbia–Adair County Volunteer Fire Department, Kentucky

Firefighter Sparks and the members of his fire department responded to a report of a fire in a residence. Once on scene, Firefighter Sparks donned full structural protective clothing and an SCBA. He entered the structure and assisted with firefighting operations. When his air supply was depleted, he exited the structure.

Firefighter Sparks replaced his air cylinder with a full one, climbed a 14-foot straight ladder to a second story window, and entered the structure to assist with salvage and overhaul of the fire. After being inside for a few moments, Firefighter Sparks told other firefighters that he was ill and collapsed. He was brought to the exterior by firefighters and provided with treatment, including the use of an AED.

He was transported to a local hospital by ambulance and later transferred to a regional hospital by air ambulance. Firefighter Sparks died as the result of a heart attack on July 8, 2011.

For additional information regarding this incident, please refer to NIOSH Fire Fighter Fatality Investigation and Prevention Program report F2011-16 (www.cdc.gov/niosh/fire/reports/face201116.html).

July 7, 2011–1550 hours
Caleb Nathanael Hamm, Firefighter
Age 23, Wildland Part-Time • Department of Interior Bureau of Land Management–Bonneville Hot Shots, Utah

Firefighter Hamm and his Hot Shot crew were assigned to the CR 377 Fire in Mineral Wells, Texas. He and his crew began their workday at 0900 hours. The day was expected to be very hot with temperatures reaching 105 °F. Firefighter Hamm and his crew were assigned to construct a fire line, and they did so until they broke for lunch at approximately 1300 hours.

After lunch, the crew broke up into squads and got back to work. At about 1550 hours, Firefighter Hamm stumbled on a slope. His partner asked him if he was okay, and Firefighter Hamm reported that he was hot and had a slight headache. His partner told him to sit down in the shade and take a rest. The other firefighter momentarily walked away. When he returned, he found Firefighter Hamm unconscious.

Wildland EMS responders were summoned, and Firefighter Hamm's condition worsened. Efforts included high altitude water drops from a helicopter on the scene to cool Firefighter Hamm and other firefighters. He was eventually transported by ambulance to the hospital, arriving at 1658 hours. He was pronounced dead at 1703 hours. The cause of death was hyperthermia.

July 11, 2011–1545 hours
John Joseph Lackovic, Jr., Fire Police Lieutenant
Age 60, Volunteer • Valley Forge Volunteer Fire Company, Pennsylvania

Fire Police Lieutenant Lackovic and the members of his department responded to a report of a motorcycle accident in their community. After approximately 10 minutes of checking with nothing found, the response was canceled.

Fire Police Lieutenant Lackovic returned home and complained of chest tightness to his daughter. He refused requests from his children to go to the hospital and subsequently collapsed at approximately 1545 hours. His daughter performed CPR until EMS responders arrived. He was transported to the hospital but did not recover. His death was caused by a heart attack.

July 17, 2011–Time Unknown
Timothy R. White, Firefighter/EMT
Age 50, Volunteer • Cedar Lake Fire Department, Indiana

Firefighter/EMT White suffered a heart attack during technical rescue training on July 17, 2011. He remained hospitalized until his death on August 5, 2011.

July 19, 2011–Time Unknown
Travis Lee Miller, Firefighter
Age 31, Volunteer • Waterloo–Grant Township Fire Department, Indiana

Firefighter Miller and the members of his fire department responded to a mutual-aid structure fire. The incident occurred at approximately 1847 hours. After the fire, Firefighter Miller returned home. He was discovered unconscious at approximately 0430 hours the next morning, the victim of a heart attack.

July 24, 2011–Time Unknown
Deon Jason "Dino" Classay, Firefighter
Age 43, Wildland Part Time • Bureau of Indian Affairs, Fort Apache Agency, Arizona

The Fort Apache Hotshot Crew was assigned to the Diamond Fire on the Fort Apache Indian Reservation near Whiteriver, Arizona. The fire was caused by lightening. Due to the remote location of the fire, firefighters were ferried into the scene by helicopter.

When fire containment efforts were complete, firefighters gathered at a helispot to await transport. A mechanical issue prevented helicopter operations. A plan was developed for firefighters to hike to a roadway where they would be met by crew carriers. Firefighters were grouped into squads for the hike up steep, heavily wooded terrain. As firefighters were accounted for at the roadway, Firefighter Classay was determined to be missing.

A search was conducted by firefighters and law enforcement personnel. The search continued through the evening and into the next day. Firefighter Classay was found deceased at 0618 hours. The cause of death has not been publically released.

July 25, 2011–Time Unknown

Gaston Aloysius Gagne, III, Firefighter

Age 46, Career • Baytown Fire Department, Texas

Firefighter Gagne passed away from an apparent heart attack after returning home from responding to three incidents earlier that day.

July 28, 2011–1230 hours

Jeffrey Scott Bowen, Captain

Age 37, Career • Asheville Fire Department, North Carolina

Asheville Fire Department units were dispatched to an automatic fire alarm activation reporting smoke on the fifth floor of a highrise medical office building. Arriving firefighters found a working fire in Suite 500 of the building. Additional fire department resources were requested. Captain Bowen was in command of Rescue 3, a unit assigned on the full structure fire dispatch.

Captain Bowen and another firefighter were conducting a search of the fifth floor. Fire department operations were complicated by locked doors, extreme heat, low visibility, and difficulties with the building's standpipe system. Captain Bowen called a Mayday when he and his firefighter were unable to exit the hazardous area. An additional alarm was called and a Rapid Intervention Crew (RIC) was deployed. Captain Bowen and his firefighter were not immediately located.

Once located, Captain Bowen was removed from the building and provided with emergency medical treatment. Despite these efforts, he died as a result of smoke inhalation.

The fire was determined to be incendiary, set by someone in Suite 500.

July 29, 2011–1248 hours
Timothy Oliveira, Lieutenant
Age 53, Career • Salisbury Fire Rescue, Massachusetts

Lieutenant Oliveira was responsible for vehicle maintenance and fleet management for Salisbury Fire Rescue. He was making preparations to change the oil on a fire department SUV in the parking lot behind the fire station.

Apparently, Lieutenant Oliveira lifted the vehicle with a portable hydraulic floor jack in preparation to perform the oil change. He used a creeper to get under the vehicle. At some point, the floor jack failed and the vehicle was lowered on top of Lieutenant Oliveira. Firefighters came outside to check on Lieutenant Oliveira and found him pinned and unconscious.

The vehicle was lifted again utilizing the floor jack. The jack failed again. Firefighters used a hydraulic rescue tool to lift the vehicle off of Lieutenant Oliveira. He was removed from under the vehicle and transported to the hospital by ambulance. He later died at the hospital. The cause of death was listed as asphyxiation due to compression of the torso.

For additional information regarding this incident, please refer to NIOSH Fire Fighter Fatality Investigation and Prevention Program report F2011-19 (www.cdc.gov/niosh/fire/reports/face201119.html).

July 29, 2011–1735 hours
Kyle Kenneth King, Captain
Age 53, Career • Perry Fire Department, Oklahoma

At 1735 hours, the Perry Fire Department was paged for a grass fire/abandoned structure fire. Captain King and all Perry Fire Department members were paged to respond through an all-call.

At 1736 hours, a local weather station reported a temperature of 104.7 °F.

When first-responding firefighters arrived on the scene, they found a grass fire that had spread to an abandoned house. Captain King and another fire officer responded in Engine 6 to the scene. They arrived at 1758 hours and immediately began to attack the structure fire. Captain King and the other fire officer advanced a 100-foot, 1-3/4 inch attack line. Both firefighters were wearing full structural firefighting protective clothing and SCBA.

The fire was called under control at 1822 hours. Captain King and other firefighters continued to work to overhaul the structure and extinguish hot spots. A short time later, all firefighters on the scene gathered at the rear of the structure to take a break and hydrate. Firefighters had taken off their coats to cool off and drink some fluids. After about 10 minutes, firefighters began to put their protective clothing on to go back to work. As Captain King began to put his helmet on, he collapsed face first to the ground.

Firefighters began to provide immediate medical assistance. Captain King was found without a pulse and not breathing. An ambulance and medical helicopter were requested. CPR was begun.

Continued on next page.

The ambulance was dispatched at 1825 hours and arrived on scene at 1840 hours. Captain King was immediately loaded in the ambulance, and an AED was attached. Captain King was transported to a local hospital. Due to his condition, Captain King was then transported by helicopter to a regional hospital in Oklahoma City. Captain King never regained consciousness and died in the hospital on August 7, 2011.

For additional information regarding this incident, please refer to NIOSH Fire Fighter Fatality Investigation and Prevention Program report F2011-24 (www.cdc.gov/niosh/fire/reports/face201124.html).

August 1, 2011–1300 hours
Jeffery Alan Cocke, Firefighter
Age 59, Volunteer • Altavista Fire Department, Virginia

Firefighter Cocke and the members of his fire department responded to two emergency incidents on August 1, 2011. At the first incident, Firefighter Cocke was the driver of a brush truck that helped to extinguish a brush fire. The incident concluded at 1540 hours. The second incident was a report of smoke in a residence. Firefighter Cocke responded to the fire station where he stood by until the incident was concluded at 1922 hours.

At approximately 1330 hours on August 2, 2011, Firefighter Cocke was transported to the hospital for difficulty breathing and a rapid heart rate. He was assessed at the hospital and released later that day. On August 4, 2011, at 0830 hours, Firefighter Cocke experienced the same symptoms. He went to the hospital and was pronounced dead a short time later. The cause of death was a pulmonary embolism.

August 11, 2011–1500 hours
Trampus S. Haskvitz, Firefighter
Age 23, Wildland Part Time • South Dakota Wildland Fire Suppression Division, South Dakota

Firefighter Haskvitz was working on State Engine 561 at the Coal Canyon Fire in Fall River County. Engine 561 was assigned fire suppression duties on a midslope road along the fire edge. The fire rapidly advanced, cutting off the escape route for the firefighters.

Two firefighters were able to run to safety, sustaining moderate to severe burn injuries. Firefighter Haskvitz could not escape from the cab of the engine and died due to smoke inhalation and thermal burns.

August 12, 2011–1215 hours
Larry Gale Nelson, Lieutenant
Age 60, Volunteer • Val Verde County Rural Volunteer Fire Department, Texas

Lieutenant Nelson was attempting to remove a metal sign from an exterior wall at Val Verde Rural Volunteer Fire Department Station Number 3. He was on a 6-foot ladder when he fell and struck his head. He suffered a fatal injury and was pronounced dead at the scene.

August 14, 2011–1612 hours
Todd Wesley Krodle, Lieutenant
Age 41, Career • Dallas Fire Rescue, Texas

Lieutenant Krodle and his ladder crew were dispatched to a fire in an apartment building. Firefighters found a working fire upon their arrival. Lieutenant Krodle and his crew were ordered to vent the roof.

Lieutenant Krodle and another firefighter accessed the roof over a ground ladder and prepared to make their first cut. Lieutenant Krodle was walking toward the ridge when the roof structure failed, and he dropped into the attic. The firefighter that was on the roof with Lieutenant Krodle attempted to reach into the hole with tools and with his hand to grab Lieutenant Krodle but was unable to do so. A Mayday was called, and firefighters attempted to locate Lieutenant Krodle from the apartments below.

Firefighters found Lieutenant Krodle part way through the ceiling of an apartment. He was hung up on structural elements and wire. Firefighters attempted to pull him through but were unable to do so. A chain saw was used to cut wooden joists, freeing Lieutenant Krodle. He was removed from the structure and provided with emergency medical treatment.

Despite the efforts of firefighters, Lieutenant Krodle died as the result of smoke inhalation.

August 16, 2011–0300 hours
Dennis James Cauthen, Fire Chief

Age 54, Volunteer • Elgin Fire Department, Lancaster, South Carolina

Chief Cauthen and the members of his fire department responded to a fire in a structure at approximately 0300 hours on August 16, 2011. Chief Cauthen drove an engine to the scene and operated the pump at the fire scene.

While at the fire scene, Chief Cauthen complained of heartburn-like symptoms. While returning to quarters, firefighters stopped to get him something to drink in hopes of relieving his heartburn. When firefighters arrived back at the fire station, Chief Cauthen's condition worsened. He was transported to the hospital but later died. The cause of his death was a heart attack.

August 26, 2011–0730 hours
Stephen Robert Cox, Fire Marshal

Age 55, Career • South Davis Metro Fire Agency, Utah

Fire Marshal Cox was at Fire Station 81 performing a mandatory annual firefighter skills fitness test. Toward the end of the skills test, Fire Marshal Cox was unable to stand on his own. Test evaluators stopped the test and sat him down in a chair, then at his request allowed him to lie down on the floor.

Crews started medical treatment due to his breathing patterns and skin color. Fire Marshal Cox started to go unconscious. Firefighters placed him in an ambulance which was located at the fire station and transported him to a local hospital. While en route to the hospital, Fire Marshal Cox suffered a heart attack. Fire Marshal Cox was treated in the emergency room then placed in the intensive care unit.

On August 27, 2011, due to his declining health, the hospital decided to transport Fire Marshal Cox to a regional hospital for further medical care. Despite these steps, Fire Marshal Cox died as a result of complications of a heart attack on August 28, 2011.

September 2, 2011–2323 hours
Henry "Jay" Branscum, Firefighter
Age 32, Volunteer • Northeast R-4 Rural Fire Association, Missouri

Firefighter Branscum was working at the scene of a mutual-aid structure fire incident involving multiple structures.

When suppression operations were complete, Firefighter Branscum was directed to assist with overhaul. These tasks involved pulling large sheets of metal to access areas of hidden fire to complete extinguishment. When overhaul was completed, Firefighter Branscum walked back uphill to his apparatus and stood by.

Approximately 15 minutes later, Firefighter Branscum collapsed. He was transported to the hospital by ambulance but later died. His death was caused by a heart attack.

September 3, 2011–1520 hours
Christopher Joseph Peterson, Firefighter
Age 22, Volunteer • Ward Four Fire Protection District, Louisiana

Firefighter Peterson was responding in the department's 1998 Dodge brush truck to an emergency medical incident. He lost control of the vehicle in a curve and was involved in a single vehicle crash. The vehicle exited the left side of the roadway and crashed into an embankment and trees.

Firefighters responded and extricated Firefighter Peterson from the vehicle. He was transported to the hospital but later died from his injuries.

News reports citing the law enforcement crash report noted that wet roads and speed were factors in the crash. The status of Firefighter Peterson's seatbelt was not known.

September 15, 2011–0800 hours
Jacob Paul Waldner, Firefighter
Age 20, Volunteer • Sunset Fire Department, Britton, South Dakota

William George Waldner, Firefighter
Age 22, Volunteer • Sunset Fire Department, Britton, South Dakota

Sunset Fire Department members were fighting a fire in a large coal bin in their community. The bin was about 75 feet tall. Firefighters had been working for some time to remove coal from the bin and believed that the incident was under control. About 80 tons of coal had already been removed with approximately 40 tons remaining in the bin.

Firefighter Jacob Waldner and Firefighter William Waldner, cousins, were on top of the bin flowing water into the bin to control the fire when an explosion occurred. The explosion ripped off the roof and one wall of the bin. Both firefighters were trapped in the debris. Both firefighters were killed.

The State Fire Marshal's report determined that the explosion was caused by spontaneous combustion. News reports explained that the type of coal in use, Powder River Basin coal, was a factor in the blast.

September 18, 2011–1850 hours
Michael Benjamin Collins, Firefighter
Age 41, Volunteer • Shelby Fire & Rescue Department, Iowa

Firefighter Collins was directing traffic around a vehicle crash. He was wearing a coat or jacket with some reflective material. A vehicle approaching the scene was going too fast, made a last minute lane change, and struck Firefighter Collins. He was pronounced dead at the scene.

September 22, 2011–Time Unknown
George Wendell Fisher, III, Captain
Age 57, Volunteer • Sandy Bottom Volunteer Fire and Rescue, North Carolina

Captain Fisher died as the result of a heart attack that he suffered within 24 hours of responding to two fire department emergency incidents.

September 25, 2011–1035 hours
Keith Gregory Rankin, Lieutenant

Age 38, Volunteer • Lancaster Township Fire Department, Pennsylvania

Lieutenant Rankin and the members of his fire department were participating in live-fire training exercises at a local training center. Lieutenant Rankin supervised the exercises and was wearing full structural firefighting protective clothing and an SCBA.

Lieutenant Rankin was talking with other firefighters when he suddenly collapsed without warning. Firefighters initiated CPR, and Lieutenant Rankin was transported to the hospital where he later died. The cause of death was a heart attack.

September 29, 2011–0636 hours
Vincent Junior Cruz, Lieutenant

Age 40, Career • Guahan Fire Department, Guam

Lieutenant Cruz was on duty in his fire station. He had not been feeling well during the shift but declined offers to go off sick. During the shift, Lieutenant Cruz responded to 13 emergency ambulance incidents, the last ending at 0139 hours.

He was discovered by other firefighters unresponsive in bed. Firefighters made efforts to revive him in the fire station and during transport to the hospital by fire department ambulance. He was pronounced dead in the hospital emergency room at 0712 hours. His death was caused by a heart attack.

October 3, 2011–1545 hours
Andrew K. "Andy" Boyt, Lieutenant
Age 45, Career • Cape May Fire Department, New Jersey

Lieutenant Boyt was on duty from 0730 hours on October 2 through 0730 hours on October 3, 2011.

The morning was taken up with administrative and maintenance duties in the fire station. Lieutenant Boyt responded on two emergency incidents during his shift.

The first incident was for an automatic fire alarm activation at a local hotel. The incident was dispatched at 1231 hours. The Avondale by the Sea Hotel is a three-story occupancy with exterior stairways. The building is not equipped with an elevator. Upon their arrival at the scene, Lieutenant Boyt and his crew donned full structural protective clothing, including SCBA, and searched all three floors of the building for the source of the alarm.

After searching the building and the area of the fire alarm activation, no source for the alarm was located, and the alarm panel was reset. Lieutenant Boyt and his crew were on scene for approximately 30 minutes. After the fire alarm panel was reset, Lieutenant Boyt and his crew returned to quarters.

At 1545 hours, Lieutenant Boyt and his crew were dispatched to a fire alarm indication in a private home. Upon their arrival, the crew found no one home and no viable way of accessing the interior of the structure to check for fire. All of the first floor entrances and windows were locked, so Lieutenant Boyt and his crew utilized portable fire ladders to access the second floor windows.

Lieutenant Boyt raised a 24-foot extension ladder to the second floor of the house and climbed the ladder to look for an open window. This process was repeated by Lieutenant Boyt for at least three windows on the residence. No open windows were located, and Lieutenant Boyt did not see any sign of fire through the windows of the house. As firefighters were leaving the scene, a key holder arrived and allowed firefighters to access the interior of the house. No fire was found, but Lieutenant Boyt used a step ladder to replace a smoke detector on the second floor of the home. The incident was concluded at 1635 hours.

There were no more emergency responses prior to the end of Lieutenant Boyt's shift. His shift ended at 0730 hours, and he departed from the fire station.

At approximately 1545 hours, Lieutenant Boyt was observed slumped over the wheel of his car. A bystander performed CPR, and Lieutenant Boyt was transported to the hospital. He did not recover.

October 22, 2011–0730 hours
Horace Christopher "Chris" Pendergrass, Firefighter
Age 49, Career • Fairfax County Fire & Rescue Department, Virginia

Firefighter Pendergrass was on duty at his fire station. He arrived on duty the morning of October 21, 2011. He worked a full shift and went to bed that evening without any health complaints to other firefighters.

At shift change the next day, Firefighter Pendergrass failed to rise from bed. Other firefighters checked his bunk room and found him unresponsive and deceased. His death was caused by a heart attack.

October 28, 2011–0734 hours
Charolette Rae "Charlie" Adair, Firefighter/EMT
Age 45, Volunteer • Richfield Township Fire Department, Ohio

Firefighter/EMT Adair was driving to work. She came upon a traffic crash within her response area and stopped to assist. She exited her vehicle and assessed the injuries of those involved in the crash. Finding no injuries, she called 9-1-1 to request law enforcement, donned a traffic vest, and began to direct traffic around the scene.

As she directed traffic, Firefighter/EMT Adair was struck by a vehicle. Fire and EMS responders were dispatched. Firefighter/EMT Adair was transported to the hospital by medical helicopter. She was pronounced dead shortly after her arrival due to traumatic injuries.

November 10, 2011–1030 hours
Edward N. Steffy, Fire Police Officer
Age 71, Volunteer • Rothsville Volunteer Fire Company, Pennsylvania

Fire Police Officer Steffy and the members of his fire department responded to a vehicle crash in their community. As firefighters completed their work at the scene, they received word that a Fire Police Officer had become ill at his traffic control position.

Firefighters responded to the traffic control position and found Fire Police Officer Steffy slumped over on the seat of his car. CPR was initiated by firefighters and bystanders. An ambulance arrived, and an AED was applied. CPR was continued in the ambulance while en route to the hospital. Fire Police Officer Steffy was pronounced dead at the hospital due to a heart attack.

November 16, 2011–1145 hours
Jonathan C.W. Young, Fire Captain
Age 49, Volunteer • Roselle Fire Department, New Jersey

In addition to serving as a volunteer firefighter, Captain Young was the emergency management coordinator for Roselle. His fire department was dispatched to a fire incident. As he drove his personal vehicle from emergency management headquarters to the fire station in response to the call, he suffered a heart attack. His vehicle crashed into a fence and a tree.

First responders arrived on the scene and provided CPR. He was transported to the hospital but was pronounced dead shortly thereafter.

November 20, 2011–0230 hours
Gregory S. Baker, Fire Chief

Age 52, Volunteer • Lewisville Community Volunteer Fire Department, Ohio

Chief Baker and the members of his fire department were dispatched at 0055 hours for a structure fire. Chief Baker worked at a tanker fill site approximately 1/4 mile from the fire scene.

At approximately 0230 hours, a firefighter working with Chief Baker contacted the IC to report that he was not feeling well. EMS responders were sent immediately to check on Chief Baker. Upon their assessment, they contacted dispatch to call for a medical helicopter. EMS responders determined that Chief Baker was having a heart attack. He was provided with ALS and defibrillated multiple times.

Chief Baker was flown by medical helicopter to a local hospital where he was later pronounced dead.

November 22, 2011–1937 hours
Johnny Lynn Norton, Firefighter

Age 56, Volunteer • Hot Springs Volunteer Fire Department, North Carolina

Firefighter Norton participated in an extended search for a lost hiker on November 21, 2011. During the incident, Firefighter Norton walked trails and roadways in remote areas and on steep grades for several hours. The incident concluded at approximately 2119 hours when the hiker was located and led to safety.

The next day, Firefighter Norton responded to an incident at a local railroad crossing. Railroad officials were on scene, and they were addressing the issue. Firefighters were released from the incident at 1813 hours.

At 1937 hours, firefighters and EMS responders were dispatched to the home of Firefighter Norton. He was found unresponsive, and firefighters began CPR. He was transported to the hospital but later pronounced dead. His death was caused by a heart attack.

December 3, 2011–0419 hours
Scott Osenenko, Firefighter

Age 45, Volunteer • Livingston Parish Fire Protection District #4, Louisiana

Firefighter Osenenko was on the first apparatus to arrive at the scene of a fire in a manufactured home. He worked on the fire scene for approximately 20 minutes assisting with rescue and firefighting activities.

While still on scene, Firefighter Osenenko began to experience chest pains. He was treated and taken to the hospital where he later died as the result of a heart attack.

December 4, 2011–0055 hours
Joseph A. "Joey" King, Firefighter
Age 60, Volunteer • Davis Creek-Ruthdale Volunteer Fire Department, West Virginia

Firefighter King and the members of his fire department responded to a report of a pile of more than 100 railroad ties on fire. Fog and smoke obscured visibility in the area. Crews could smell smoke in the area but could not find the fire.

Firefighter King was on a local bridge looking for the fire when he fell from the bridge to the embankment below. He was killed in the fall.

The fire was determined to be incendiary. The alleged arsonist was charged with murder for the death of Firefighter King.

December 8, 2011–0243 hours
Kevin Edward Townes, Sr., Firefighter
Age 54, Career • Mount Vernon Fire Department, New York

Firefighter Townes was on duty for a 24-hour shift that began at 0800 hours on December 7, 2011. During the first part of the shift, Firefighter Townes responded to four emergency incidents. The fifth incident, a structure fire, was dispatched at 0243 hours.

Firefighter Townes responded to the scene on Engine 4. When he arrived on the scene, he donned his personal protective clothing and SCBA and prepared to enter the structure when he suddenly collapsed.

Firefighter Townes was treated on the scene by firefighters and paramedics. He was transported to the hospital by ambulance. He was treated in the emergency room but was not revived. He was pronounced dead at 0406 hours due to a heart attack.

December 8, 2011–0418 hours
Jon D. Davies, Sr., Firefighter
Age 43, Career • Worcester Fire Department, Massachusetts

Firefighter Davies was assigned to Rescue 1. At 0418 hours, his unit and other units of the Worcester Fire Department were dispatched to a report of a fire in a three-story wood frame residence. Firefighters found a working fire to the rear of the structure and went to work.

Responding to reports of a civilian trapped on the third floor, Firefighter Davies and the members of his crew conducted a primary search. After the search was completed, the IC ordered all firefighters from the building and implemented defensive firefighting operations. At 0443 hours, a partial collapse of the roof and rear porches occurred.

A resident of the building continued to insist that his roommate was still trapped in the building. The fire had been knocked down, and it was determined that another search of the structure would be conducted in hopes of locating the missing roommate. Approximately 4 minutes into the search, a structural collapse occurred and trapped two members of Rescue 1, including Firefighter Davies. He was crushed and fatally injured in the collapse. The other firefighter was seriously injured but was rescued from the building.

December 23, 2011–1324 hours
James M. "Jim" Rice, Firefighter
Age 42, Career • Peabody Fire Department, Massachusetts

Firefighter Rice and the members of his fire department responded to a fire in an apartment building. Firefighter Rice, wearing full structural protective clothing and an SCBA, entered the structure to fight the fire. Firefighter Rice collapsed while working on the second floor of the building.

He was removed from the structure by firefighters and treated at the scene. He was rushed to the hospital by ambulance but did not survive. The nature of Firefighter Rice's fatal injuries have not been disclosed.

Firefighter Fatalities from Previous Years

February 5, 1978–Time Unknown
Thomas "Todd" Lange, Firefighter

Age at Injury 26, Age at Death 60, Career • Pittsfield Fire Department, Massachusetts

Firefighter Lange was performing mouth-to-mouth resuscitation on a patient on February 5, 1978. The patient vomited directly into Firefighter Lange's mouth during treatment. The fire department was notified by the hospital that the patient was a hepatitis carrier. Firefighter Lange was diagnosed with Hepatitis C in 1997 and retired as a result of his disease in 2006. He died on May 20, 2011, and his death certificate listed the cause of death as Hepatitis C due to cirrhosis due to metastatic hepatocellular disease.

June 19, 1993–Time Unknown
Harold Gregory Gibson, Firefighter III

Age at Injury 35, Age at Death 53, Career • Fulton County Fire Rescue, Georgia

Firefighter Gibson contracted Hepatitis C after coming in contact with blood while treating a patient injured in a traffic crash in 1993. Due to medical complications related to Hepatitis, Firefighter Gibson left the fire department in 1998 and died on July 2, 2011.

September 7, 2010–Time Unknown
Donald R. Lam, Jr., Forest Ranger Technician III

Age 58 • Kentucky Division of Forestry, Kentucky

Forest Ranger Technician III Lam was working on the scene of a wildland fire in Livingston County, Kentucky. As he and other firefighters constructed a fire line at the base of a bluff, a dead tree on the top of the bluff fell and rolled over the bluff.

Forest Ranger Technician III Lam was struck in the head by the tree. He was treated at the scene and transported to the hospital. He was released from the hospital briefly from November 2010 to January 2011. Forest Ranger Technician III Lam died on February 17, 2011, due to complications from his injuries.

Appendix B

Firefighter Fatality Inclusion Criteria – National Fire Service Organizations

The National Fire Protection Association (NFPA), the National Fallen Firefighters Foundation (NFFF), the U.S. Fire Administration (USFA), and other organizations individually collect information on firefighter fatalities in the United States. Each organization uses a slightly different set of inclusion criteria that are based at least in part on the purposes of the information collection for each organization and data consistency.

As a result of these differing inclusion criteria, statistics about firefighter fatalities may be provided by each organization that do not coincide with one another. This section will explain the inclusion criteria for each organization and provide information about these differences.

The USFA includes firefighters in this report who died while on duty, became ill while on duty and later died, and firefighters who died within 24 hours of an emergency response or training regardless of whether the firefighter complained of illness while on duty. The USFA counts firefighter deaths that occur in the 50 States, the District of Columbia, and United States protectorates, such as Puerto Rico and Guam. Detailed inclusion criteria for this report appear starting on page 66 of this report.

For 2011, the USFA reported 83 onduty firefighter fatalities.

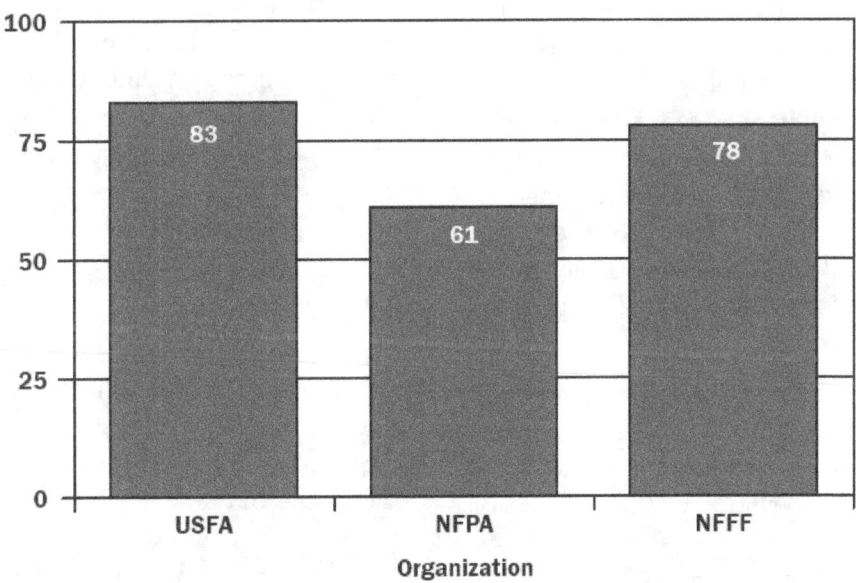

Inclusion Criteria for NFPA's Annual Firefighter Fatality Study

Introduction

Each year, NFPA collects data on all firefighter fatalities in the United States that resulted from injuries or illnesses that occurred while the victims were on duty. The purpose of the study is to analyze trends in the types of illnesses and injuries resulting in deaths that occur while firefighters are on the job. This annual census of firefighter fatalities in its current format dates back to 1977. (Between 1974 and 1976, NFPA published a study of onduty firefighter fatalities that was not as comprehensive.)

What is a Firefighter?

For the purpose of the NFPA study, the term *firefighter* covers all uniformed members of organized fire departments, whether career, volunteer or combination, or contract; full-time public service officers acting as firefighters; State and Federal government fire service personnel; temporary fire suppression personnel operating under official auspices of one of the above; and privately employed firefighters including trained members of industrial or institutional fire brigades, whether full or part time.

Under this definition, the study includes, besides uniformed members of local career and volunteer fire departments, those seasonal and full-time employees of State and Federal agencies who have fire suppression responsibilities as part of their job description, prison inmates serving on firefighting crews, military personnel performing assigned fire suppression activities, civilian firefighters working at military installations, and members of industrial fire brigades. Impressed civilians would also be included if called on by the officer in charge of the incident to carry out specific duties. The NFPA study includes fatalities that occur in the 50 States and the District of Columbia.

What does "on duty" mean?

The term *on duty* refers to being at the scene of an alarm, whether a fire or nonfire incident; being en route while responding to or returning from an alarm; performing other assigned duties such as training, maintenance, public education, inspection, investigations, court testimony, and fundraising; and being on call, under orders, or on standby duty other than at home or at the individual's place of business. Fatalities that occur at a firefighter's home may be counted if the actions of the firefighter at the time of injury involved firefighting or rescue.

Onduty fatalities include any injury sustained in the line of duty that proves fatal, any illness that was incurred as a result of actions while on duty that proves fatal, and fatal mishaps involving non-emergency occupational hazards that occur while on duty. The types of injuries included in the first category are mainly those that occur at an incident scene, in training, or in accidents while responding to or returning from alarms. Illnesses (including heart attacks) are included when the exposure or onset of symptoms are tied to a specific incident of onduty activity. Those symptoms must have been in evidence while the victim was on duty for the fatality to be included in the study.

Fatal injuries and illnesses are included even in cases where death is considerably delayed. When the onset of the condition and the death occur in different years, the incident is counted in the year of the condition's onset. Medical documentation specifically tying the death to the specific injury is required for inclusion of these cases in the study.

Categories not included in the study

The NFPA study does not include members of fire department auxiliaries; non-uniformed employees of fire departments; emergency medical technicians (EMTs) who are not also firefighters; chaplains; or civilian dispatchers. The study also does not include suicides as onduty fatalities even when the suicide occurs on fire department property.

The NFPA recognizes that a comprehensive study of firefighter onduty fatalities would include chronic illnesses (such as cardiovascular disease and certain cancers) that prove fatal and that arose from occupational factors. In practice, there is as yet no mechanism for identifying onduty fatalities that are due to illnesses that develop over long periods of time. This creates an incomplete picture when comparing occupational illnesses to other factors as causes of firefighter deaths. This is recognized as a gap the size of which cannot be identified at this time because of the limitations in tracking the exposure of firefighters to toxic environments and substances and the potential long-term effects of such exposures.

2011 Experience

In 2011, a total of 61 onduty firefighter deaths occurred in the United States, according to the NFPA inclusion criteria.

National Fallen Firefighters Foundation

In 1997, fire service leaders formulated new criteria to determine eligibility for inclusion on the National Fallen Firefighter Memorial. Line-of-duty deaths (LODDs) shall be determined by the following standards:

1. (a) Deaths of firefighters meeting the Department of Justice's (DOJ's) Public Safety Officers' Benefits (PSOB) program guidelines, and those cases that appear to meet these guidelines whether or not PSOB staff has adjudicated the specific case prior to the annual National Fallen Firefighters Memorial Service; and

   (b) Deaths of firefighters from injuries, heart attacks, or illnesses documented to show a direct link to a specific emergency incident or department-mandated training activity.

2. While PSOB guidelines cover only public safety officers, the Foundation's criteria also include contract firefighters and firefighters employed by a private company, such as those in an industrial brigade, provided that the deaths meet the standards listed above.

3. Some specific cases will be excluded from consideration, such as deaths attributable to suicide, alcohol or substance abuse, or other gross abuses as specified in the PSOB guidelines.

The National Fallen Firefighters Memorial was built in 1981 in Emmitsburg, MD. The names listed there begin with those firefighters who died in the line of duty that year. The U.S. Congress created the National Fallen Firefighters Foundation to lead a nationwide effort to remember America's fallen firefighters. Since 1992, the tax-exempt, nonprofit foundation has developed and expanded programs to honor our fallen fire heroes and assist their families and coworkers by providing them with resources to rebuild their lives. Since 1997, the Foundation has managed the National Memorial Service held each October to honor the firefighters who died in the line of duty the previous year.

As of this writing, the Foundation will be honoring 81 firefighters who died in the line of duty at the October 2012 Memorial Weekend. Of those 81 being honored, 78 died in 2011 as the result of incidents that occurred in 2011, and three others died in previous years as the result of incidents that occurred in previous years.

The following section is a listing of the firefighters who will be honored by the Foundation in October of 2012 as of this writing.

Firefighter deaths that occurred in 2011 as the result of incidents that occurred in 2011:

Arkansas

David L. Eason
West Memphis Fire Department

California

Glenn L. Allen
Los Angeles Fire Department

William F. Hopman
Quincy Volunteer Fire Department

Vincent A. Perez
San Francisco Fire Department

Anthony M. Valerio
San Francisco Fire Department

Connecticut

Robert D. Watts
Windsor Volunteer Fire Department

Florida

Joshua O. Burch
Florida Forest Service

Brett L. Fulton
Florida Forest Service

Gregory L. Harris
Miami Dade County Fire Rescue Department

James L. von Roden
Lee Community Volunteer Fire Department

Guam

Vincent J. Cruz
Guam Fire Department

Idaho

Caleb N. Hamm
Bureau of Land Management, Bonneville
Interagency Hotshot Crew

Illinois

Daniel C. Dare
Avon Fire Protection District

Patrick B. Hannon
Chicago Fire Department

Corey R. Shaw
Du Quoin Fire Department

Indiana

Steven F. Auch
Indianapolis Fire Department

Scott T. Davis
Muncie Fire Department

Travis L. Miller
Waterloo-Grant Township Volunteer
Fire Department

Timothy R. White
Cedar Lake Fire Department

Kansas

Jim D. Niles
Downs Fire Department

Kentucky

Donald R. Lam
Kentucky Division of Forestry

Charles V. Sparks
Columbia-Adair County Volunteer
Fire Department

Christopher T. Stock
Westport Volunteer Fire Department

Michael C. Webb
Fleming-Neon Volunteer Fire Department

Louisiana

Scott Osenenko
Livingston Parish Fire Protection District #4

Christopher Peterson
Ward Four Fire Protection District

Maine

David E. Remington Sr.
Shapleigh Fire & Rescue Department

Maryland

Mark G. Falkenhan
Baltimore County Fire Department,
Lutherville Volunteer Fire Company

Massachusetts

Jon D. Davies Sr.
Worcester Fire Department

James M. Rice
Peabody Fire Department

Minnesota

Chip A. Imker
Cambridge Fire Department

Mississippi

Jacob A. Carter
Becker-Athens Volunteer Fire Department

Charles E. Foster
Barton Fire Department

Mississippi *(continued)*

Larry C. Gressett Sr.
Toomsuba-Alamucha Fire & Rescue

Joshua L. Wilkes
Unity Fire Department

Missouri

Henry J. Branscum
Northeast R-IV Rural Fire Association

Leslie L. Clark
Dixon Rural Fire Protection District

Richard E. Paul
Kansas City Fire Department

New Hampshire

Harold F. Frey
Sandown Fire and Rescue Department

New Jersey

Andrew K. Boyt
City of Cape May Fire Department

Thomas Shields
Flanders Fire Company No. 1

Jonathan C. W. Young
Roselle Fire Department

New York

Jarrett T. Eleam
Big Tree Volunteer Fire Company

Michael P. Esposito
Baldwin Fire Department

Thomas V. Regan
Garden City Park Fire Department

Kevin E. Townes
Mount Vernon Fire Department

North Carolina

Richard L. Barbour
Wilson's Mills Fire/Rescue

Jeffrey S. Bowen
Asheville Fire Department

George W. Fisher III
Sandy Bottom Volunteer Fire and Rescue

David S. Howell
Roseboro Volunteer Fire Department

David J. Hunsinger Jr.
Tar Heel Volunteer Fire Department

Johnny L. Norton
Hot Springs Volunteer Fire Department

Ohio

Charolette R. Adair
Richland Township Fire Department

Gregory S. Baker
Lewisville Community Volunteer Fire Department

Randy D. Boley
Clinton Township Fire Department

Robert J. Tieche
Cardinal Joint Fire District

Oklahoma

Kyle K. King
Perry Fire Department

Pennsylvania

Christian D. Beaston Jr.
Hellam Fire Company

Derek Kozorosky
United States Air Force, Kadena Air Force Base, Japan

John J. Lackovic Jr.
Valley Forge Volunteer Fire Company

Keith G. Rankin
Lancaster Township Fire Department

Edward N. Steffy
Rothsville Volunteer Fire Company

South Carolina

Dennis J. Cauthen
Elgin Volunteer Fire Department

Robin E. West
Startex Fire Department

South Dakota

Trampus S. Haskvitz
South Dakota Department of Agriculture, Division of Wildland Fire Suppression

Jacob P. Waldner
Sunset Fire Department

William G. Waldner
Sunset Fire Department

Texas

Gaston A. Gagne III
Baytown Fire Department

Elias M. Jaquez
Cactus Volunteer Fire Department

Todd W. Krodle
Dallas Fire-Rescue Department

Chris K. Pham
Dallas Fire-Rescue Department

Gregory M. Simmons
Eastland Fire Department

Utah

Stephen R. Cox
South Davis Metro Fire Agency

Virginia

Jeffery A. Cocke
Altavista Volunteer Fire Department

Washington

Matthew M. Hadaller III
Lewis County Fire District #3

Garet G. Rasmussen
Chelan County Fire District 1

West Virginia

Joseph A. King Jr.
Davis Creek-Ruthdale Volunteer Fire Department

Wisconsin

Ronald D. Ruprecht
Stone Lake Fire Department

Deaths From Previous Years

Georgia

Terrell G. Nielsen Sr.
Bryan County Fire Department

Montana

Remy H. Pochelon
USDA Forest Service

Ohio

James M. Hall
Greentown Fire Department

www.ingramcontent.com/pod-product-compliance
Lightning Source LLC
Chambersburg PA
CBHW081219170526
45165CB00009B/2867